# Interconnections Between Human and Ecosystem Health

## Chapman & Hall Ecotoxicology Series

Series Editors

**Michael H. Depledge**
Professor of Ecotoxicology, Institute of Biology, Odense University, Denmark

**Brenda Sanders**
Associate Professor of Physiology, Molecular Ecology Institute, California State University, USA

In the last few years emphasis in the environmental sciences has shifted from direct toxic threats to humans, towards more general concerns regarding pollutant impacts on animals and plants, ecosystems and indeed on the whole biosphere. Such studies have led to the development of the scientific discipline of ecotoxicology. Throughout the world socio-political changes have resulted in increased expenditure on environmental matters. Consequently, ecotoxicological science has developed extremely rapidly, yielding new concepts and innovative techniques that have resulted in the identification of an enormous spectrum of potentially toxic agents. No single book or scientific journal has been able to keep pace with these developments.

This series of books provides detailed reviews of selected topics in ecotoxicology. Each book includes both factual information and discussions of the relevance and significance of the topic in the broader context of ecotoxicological science.

Already published

**Animal Biomarkers as Pollution Indicators**
David B. Peakall
Hardback (0 412 40200 9), 292 pages

**Ecotoxicology in Theory and Practice**
V. E. Forbes and T. L. Forbes
Hardback (0 412 43530 6), 262 pages

# Interconnections Between Human and Ecosystem Health


Edited by

**Richard T. Di Giulio**
Duke University
Durham
USA

and

**Emily Monosson**
University of Massachusetts
Department of Forestry and Wildlife Management
Amherst
USA



SPRINGER-SCIENCE+
BUSINESS MEDIA, B.V.

First edition 1996


Originally published by Chapman & Hall in 1996
Softcover reprint of the hardcover 1st edition 1996

Typeset in 10/12pt Times by WestKey Ltd, Falmouth, Cornwall

ISBN 978-94-010-7181-9 ISBN 978-94-009-1523-7 (eBook)
DOI 10.1007/978-94-009-1523-7



A catalogue record for this book is available from the British Library

Library of Congress Catalog Card Number: 95-71854

Printed on permanent acid-free text paper, manufactured in accordance with ANSI/NISO Z39.48-1992 and ANSI/NISO Z39.48-1984 (Permanence of Paper).

# Contents

*We dedicate this work to our children,*
*Margaret Ellen Di Giulio*
*and Samuel Monosson Letcher,*
*and their generation*

# *Contributors*

S. Marshall Adams
Environmental Sciences Division,
Bld 1505, PO Box 2008,
Oak Ridge National Laboratory,
Oak Ridge, TN 37831-6036

Lawrence J. Axelrod
Department of Psychology,
University of British Columbia,
Vancouver, British Columbia

Linda A. Baldwin
School of Public Health,
University of Massachusetts,
Public Health Building,
Amherst, MA 01003

Steven M. Bartell
Senes Oak Ridge Inc.,
Center for Risk Analysis,
677 Emory Valley Road, Suite C,
Oak Ridge, TN 37830

Andrew Baum
Behavioral Medicine and Oncology,
Pittsburgh Cancer Institute,
3600 Forbes Avenue,
Suite 405,
Pittsburgh, PA 15213

Joanna Burger
Department of Biological Sciences,
and Environmental and
Occupational Health Sciences
Institute,
Rutgers University,
Piscataway, NJ 08855-1059

John Cairns, Jr
Department of Biology,
Virginia Polytechnic Institute and
State University,
Blacksburg, VA 24061

Edward J. Calabrese
School of Public Health,
University of Massachusetts,
Public Health Building,
Amherst, MA 01003

Theo Colborn
World Wildlife Fund,
1250 24th St, NW,
Washington, DC 20037

Richard T. Di Giulio
School of the Environment,
Duke University,
Durham, NC 27708-0328

Terrell Dixon
Department of English,
University of Houston,
4800 Calhoun,
Houston, TX 77204-3012

Michael Gochfeld
Environmental and Community Medicine, and Environmental and Occupational Health Sciences Institute,
UMDNJ-Robert Wood Johnson Medical School,
Piscataway, NJ 08854

Cheryll Glotfelty
English Department,
University of Nevada,
Reno, NV 89557

L. Earl Gray, Jr
Reproductive Toxicology Branch,
Developmental Toxicology Division,
Health Effects Research Laboratory,
US EPA,
Research Triangle Park, NC 27711

Mark S. Greeley
Environmental Sciences Division,
Bld 1505, PO Box 2008,
Oak Ridge National Laboratory,
Oak Ridge, TN 37831-6036

Mark E. Hahn
Biology Department, Redfield-3,
Woods Hole Oceanographic Institution,
Woods Hole, MA 02543

William R. Kelce
Reproductive Toxicology Branch,
Developmental Toxicology Division,
Health Effects Research Laboratory,
US EPA,
Research Triangle Park, NC 27711

Fumio Matsumura
Toxic Substance Program,
LEHR Building,
University of California–Davis,
Davis, CA 95616

Timothy McDaniels
Westwater Research Centre,
University of British Columbia,
Vancouver, British Columbia

Emily Monosson
Department of Forestry and Wildlife Management,
University of Massachusetts,
Amherst, MA 01003

Mary K. O'Keeffe
Department of Psychology,
Providence College,
Providence, RI 02918

Paul Slovic
Decision Research,
1201 Oak Street,
Eugene, OR 97401

V. Kerry Smith
School of the Environment,
Duke University,
Durham, NC 27708-0328

Rebecca J. Van Beneden
5751 Murray Hall,
Department of Zoology,
University of Maine,
Orono, Maine 04469-5751

# *Series foreword*

Ecotoxicology is a relatively new scientific discipline. Indeed, it might be argued that it is only during the last 5–10 years that it has come to merit being regarded as a true science, rather than a collection of procedures for protecting the environment through management and monitoring of pollutant discharges into the environment. The term 'ecotoxicology' was first coined in the late sixties by Prof. Truhaut, a toxicologist who had the vision to recognize the importance of investigating the fate and effects of chemicals in ecosystems. At that time, ecotoxicology was considered a sub-discipline of medical toxicology. Subsequently, several attempts have been made to portray ecotoxicology in a more realistic light. Notably, both Moriarty (1988) and F. Ramade (1987) emphasized in their books the broad basis of ecotoxicology, encompassing chemical and radiation effects on all components of ecosystems. In doing so, they and others have shifted concern from direct chemical toxicity to humans, to the far more subtle effects that pollutant chemicals exert on natural biota. Such effects potentially threaten the existence of all life on earth.

Although I have identified the sixties as the era when ecotoxicology was first conceived as a coherent subject area, it is important to acknowledge that studies that would now be regarded as ecotoxicological are much older. Wherever people's ingenuity has led them to change the face of nature significantly, it has not escaped them that a number of biological consequences, often unfavourable, ensue. Early waste disposal and mining practices must have alerted the practitioners to effects that accumulated wastes have on local natural communities; for example, by rendering water supplies undrinkable or contaminating agricultural land with toxic mine tailings. As activities intensified with the progressive development of human civilizations, effects became even more marked, leading one early environmentalist, G. P. Marsh, to write in 1864: 'The ravages committed by Man subvert the relations and destroy the balance that nature had established'.

But what are the influences that have shaped the ecotoxicological studies of today? Stimulated by the explosion in popular environmentalism in the sixties, there followed in the seventies and eighties a tremendous increase in the creation of legislation directed at protecting the environment. Furthermore, political restructuring, especially in Europe, has led to the widespread implementation of this legislation. This currently involves enormous numbers of environmental managers, protection officers, technical staff and consultants. The ever-increasing use of new chemicals places further demands on government agencies and industries who are required by law to evaluate potential toxicity and likely environmental impacts. The environmental manager's problem is that he needs rapid answers to current questions concerning a very broad range of chemical effects and also information about how to control discharges, so that legislative targets for *in situ* chemical levels can be met. It is not surprising, therefore. that he may well feel frustrated by more research-based ecotoxicological scientists who constantly question the relevance and validity of current test procedures and the data they yield. On the other hand, research-based ecotoxicologists are often at a loss to understand why huge amounts of money and time are expended on conventional toxicity testing and monitoring programmes, which may satisfy legislative requirements, but apparently do little to protect ecosystems from long-term, insidious decline.

It is probably true to say that until recently ecotoxicology has been driven by the managerial and legislative requirements mentioned above. However, growing dissatisfaction with laboratory-based tests for the prediction of ecosystem effects has enlisted support for studying more fundamental aspects of ecotoxicology and the development of conceptual and theoretical frameworks.

Clearly, the best way ahead for ecotoxicological scientists is to make use of the strengths of our field. Few sciences have at their disposal such a well-integrated input of effort for people trained in ecology, biology, toxicology, chemistry, engineering, statistics, etc. Nor have many subjects such overwhelming support from the general public regarding our major goal: environmental protection. Equally important, the practical requirements of ecotoxicological managers are not inconsistent with the aims of more academically-orientated ecotoxicologists. For example, how better to validate and improve current test procedures than by conducting parallel basic research programmes *in situ* to see if controls on chemical discharges really do protect biotic communities?

More broadly, where are the major ecotoxicological challenges likely to occur in the future? The World Commission on Environment and Development estimates that the world population will increase from *c.* 5 billion at present to 8.2 billion by 2025. 90% of this growth will occur in developing countries in subtropical and tropical Africa, Latin America and Asia. The introduction of chemical wastes into the environment in these regions is likely to escalate dramatically, if not due to increased industrial output, then due to the use of pesticides and fertilizers in agriculture and the disposal of damaged, unwanted or obsolete consumer goods supplied from industrialized countries. It may be many years before resources become available to implement effective waste-recycling programmes in countries with poorly developed infrastructures, constantly threatened by natural disasters and poverty.

Furthermore, the fate, pathways and effects of chemicals in subtropical and tropical environments have barely begun to be addressed. Whether knowledge gained in temperate ecotoxicological studies is directly applicable in such regions remains to be seen.

The Chapman & Hall Ecotoxicology Series brings together expert opinion on the widest range of subjects within the field of ecotoxicology. The authors of the books have not only presented clear, authoritative accounts of their subject areas, but have also provided the reader with some insight into the relevance of their work in a broader perspective. The books are not intended to be comprehensive reviews, but rather accounts which contain the essential aspects of each topic for readers wanting a reliable introduction to a subject or an update in a specific field. Both conceptual and practical aspects are considered. The Series will be constantly added to and books revised to provide a truly contemporary view of ecotoxicology. I hope that the Series will prove valuable to students, academics, environmental managers, consultants, technicians, and others involved in ecotoxicological science throughout the world.

Michael Depledge
*University of Plymouth, UK*

# Acknowledgments

We are most grateful to the contributors to this book, who also provided thoughtful reviews for chapters other than their own. In addition we are indebted to the individuals who also reviewed the chapters for this book: Dr Ed Casillas, NMFS/NOAA, Seattle, WA; Dr Adria Elskus, State University of New York, Stony Brook, NY; Dr Randall Kramer, Duke University, Durham, NC; Dr Gerald LeBlanc, North Carolina State University, Raleigh, NC; Dr Tina Letcher, Kingston, RI; Dr Michael Moore, Woods Hole Oceanographic Institution, Woods Hole, MA; Dr Kenneth Reckhow, Duke University; and Ms Stella Ross, Waste Management Institute, SUNY, Stony Brook.

In addition we would like to thank those who contributed to the original SETAC panel on this topic: Dr David Hinton, University of California, Davis, CA; Dr John Stegeman, Woods Hole Oceanographic Institution; and Dr Peter Thomas, University of Texas at Austin, Port Aransas, TX.

Finally we thank others who encouraged us to pursue this topic, and who provided us with a 'sounding board' when searching for authors, ideas and connections: Dr Joseph Bonaventura, Duke University; Dr John Cairns, Virginia Tech, Blacksburg, VA; Dr Michael Dieter, NIEHS, Research Triangle Park, NC; Dr John Grupenhoff, Physicians for the Environment, Bethesda, MD; Dr Ben Letcher, Anadromous Fish Research Center, NBS, Turners Falls, MA; Mr Sonny Monosson, Boston Financial and Equity, Boston, MA; Dr Eugene Odum, University of Georgia, Athens, GA; and Dr Paul Stern, National Research Council, Washington, DC.

# *Introduction*

# 1 Interconnections between human and ecosystem health: opening lines of communication

RICHARD T. DI GIULIO AND EMILY MONOSSON

When we embarked on this project, we were motivated by concerns that are in essence two sides of the same coin. We were convinced that fundamental connections indeed exist between human and ecosystem 'health' (we will get to problems of definition later); however, we believed that these connections were being largely ignored, to the detriment of efforts to protect or manage the environment for the well-being of both humans and ecosystems. Perhaps our interest in this concept arises from our disciplinary background; we are environmental toxicologists who study mechanisms by which pollutants affect aquatic organisms. Most scientists in this field share a fascination with the commonality (or 'conservation' in the lingo of evolutionary biology) of molecular adaptations by which diverse organisms, including humans, attempt to defend themselves from environmental insults (including chemical contaminants), as well as with modes of chemical toxicity that are similar in diverse organisms. This notion of shared mechanisms comprises perhaps the most obvious level of connectedness pertinent to this undertaking. Indeed, the comparative approach has a vibrant history within biology for shedding light on various aspects of human health, including nervous system function, reproduction and development, and cancer, to name a few. Our interests (and those of all contributors to this book), however, extend far beyond this concept of animal 'models' for the human organism. We are concerned with fundamental relationships between human and ecosystem health, and ask questions such as: How does ecosystem health impinge upon human health? How do humans affect ecosystem health (since human health really does not affect ecosystems)? How might/should these relationships affect the way we go about our

*Interconnections Between Human and Ecosystem Health.* Edited by Richard T. Di Giulio and Emily Monosson. Published in 1996 by Chapman & Hall, London. ISBN 0 412 62400 1.

science(s)? Assess risks? Manage environments? Also, how do these relationships affect human culture, and vice versa?

As alluded to earlier, the flip side of our fascination with commonalities, that also provided motivation, was a frustration with the gulf that exists between scientists dealing with various aspects of human and ecosystem health. An early goal of this book was to facilitate communication among scientists working in what appeared to us as closely related yet bafflingly estranged fields of inquiry. The project evolved, as described following, to include others outside of natural science that shared a keen interest in this topic of interconnections (such as environmentally oriented social scientists, and writers of literature dealing with nature and their critics). In the process, we realized that there were more than two islands in this gulf. With this book we have brought together a diverse collection of disciplinary specialists, each highly respected in his or her field, to offer their perspectives on this subject of human/ecosystem connectedness in a manner accessible to non-specialists. Thus the goal of this book remains to open lines of communication among disciplines in order to foster crossfertilizations that will ultimately contribute to our shared, ultimate goal of protecting and enhancing the well-being of ecosystems and their human inhabitants.

In discussing the subject of this book with others (natural scientists, social scientists, writers, friends, lovers, etc.), it became clear that for most, the assumption of interconnections between human and ecosystem health is a given; it is both intuitive and significant. Yet scientists dealing with environmental pollution agreed that it is a neglected topic. Why is this the case? Perhaps in part it is because the topic is by nature interdisciplinary while modern science, particularly the educational system that trains and rewards research scientists, strongly favors specialists (see John Cairns's eloquent discussion of this issue in Chapter 8). Another problem is embedded in the term 'ecosystem health'. This term has generated considerable discussion and debate (see, for example, Costanza *et al.*, 1992; Suter, 1993). We purposely have not defined this term, in part to avoid constraining contributors to this book. And moreover, while we agree with much of Glenn Suter's criticism of the term (for example, that it implies to some that ecosystems are like 'superorganisms' that can be diagnosed and treated like patients), we continue to use it precisely because it has different shades of meaning to different people. And for most, it conjures up a powerful, albeit abstract, concept that runs the gamut for different individuals from 'superorganism' to 'sustainability'. Thus, at a time in which the concept of interconnections is in its early development, we felt that the term 'ecosystem health' was appropriate because of its flexibility and power. And besides, none of the contributors objected to it!

The development of this project reflects our own evolution (albeit nascent and slow) from specialists into fledgling inter- (or at least multi-) disciplinarians. This development is also reflected in the organization of the book. We originally focused upon biological aspects of interconnections, with consideration to taking a biological levels of organization approach (i.e. examining potential connections at the molecular, cellular, organismal, population, community and ecosystem levels). However, we were quickly drawn into a broader perspective. A key utility of the research performed by toxicologists (both human and ecotoxicologists) is to provide data for risk assessments. Indeed, risk assessment is the dominating paradigm today in regulatory and managerial aspects of environmental pollution. And perhaps counter to the spirit of this project, much debate in this area currently focuses upon fundamental differences between human and ecological risk assessments (see Burger and Gochfeld, Chapter 9). While recognizing these differences, we question whether the complete separation of these activities is in the short run efficient, and in the long run intellectually sound or socially productive. In any event, varying perspectives on the risk assessment paradigm comprise positions perhaps appropriately in the middle of the book.

Once one enters into the realm of risk assessment, socioeconomic perspectives that establish frameworks for decision-making become as important as the data provided by natural scientists. These perspectives include economic evaluations of maintaining 'healthy' ecosystems, differences in how people perceive risks to themselves versus the environment, and how perceptions of environmental degradation may impact human health. This last, psychological perspective provides a most appropriate bridge to the final section of the book. This section describes how modern concerns for environmental degradation have influenced literature, both fiction and non-fiction, and the resulting emergence of a new, vibrant component of literary criticism, 'ecocriticism'.

In going through this book, we hope the reader will come to a greater appreciation, as we surely have, for the diversity of perspectives on this issue of interconnections. And certainly, we have come away from this project with a strong sense of the significance of this topic, though often ill-defined or not stated explicitly, in the lives of many people. Again, our goal is to enhance communications among those dealing with various issues related to human health–ecosystem health connections. Our ambition here seemed relatively modest at the outset – to facilitate interactions between the biomedical and ecological scientific communities dealing with environmental pollution. As the project developed, however, a movement towards broader interactions among natural scientists, social scientists and humanists seemed to naturally evolve. In the course of this project, other related events have occurred. For example, in June of 1994, a meeting was held in Ottawa, Ontario, entitled the 'First International Symposium on Ecosystem Health and Medicine: Integrating Science, Policy and Management'. This symposium marked, to our knowledge, the first major meeting dealing with the topic of interconnections. In June, 1995, the 'First Conference of the Association for the Study of Literature and Environment' was held in Fort Collins, Colorado. Thus it appears that what has been heretofore almost a 'closet' issue is beginning to make its way into the mainstream of professional scientific and literary activities.

We sense, however, that considerable inertia will have to be overcome for this topic to be dealt with in a manner that is simultaneously multidisciplinary and in-depth. For example, the two major societies dealing with toxicology are the Society of Toxicology and the Society of Environmental Toxicology and Chemistry, which focus upon biomedical and ecological concerns, respectively. As Mark Hahn points out (Chapter 2), fewer than 200 individuals belong to both organizations which collectively have about 6500 members. In fact, each society rarely announces meetings held by the other, despite having conduits for such announcements. That these closely related fields appear to have such difficulty in interacting is telling, and may relate to the paucity of formal discussions on the subject of interconnections, at least at the level of the natural sciences. We recognize that this project really only scratches the surface of this complex topic. If it serves as a starting point for additional discussion that over time helps to build bridges between human health-oriented and ecosystem-oriented natural scientists, social scientists, policy-makers and managers, and writers, this book will have fulfilled its purpose.

## REFERENCES

Costanza, R., Norton, B. G. and Haskell, B. D. (eds) (1992) *Ecosystem Health. New Goals for Environmental Management*, Island Press, Washington, DC.

Suter, G. W. II (1993) A critique of ecosystem health concepts and indices. *Environ. Toxicol. Chem.* **12,** 1533–9.

# PART ONE

## *Mechanistic Linkages*

# 2 *Ah receptors and the mechanism of dioxin toxicity: insights from homology and phylogeny*

MARK E. HAHN

## 2.1 INTRODUCTION

The growing recognition of the interconnections between human health and the health of ecosystems is based in part on the concept of shared responses to environmental pollutants, reflecting shared mechanisms. Fundamental to this concept is an understanding of comparative biology, the unity and diversity in the structure and function of living organisms. Yet the impact of environmental contaminants on humans and non-human organisms continues to be studied, for the most part, by separate groups of scientists with limited interaction. As an illustration of this problem, one need only look at the membership of the premier scientific societies devoted to toxicology. The Society of Toxicology, whose primary focus is on human health, has approximately 3600 members. The Society of Environmental Toxicology and Chemistry, which focuses on the non-human components of ecosystems, has approximately 2900 members. Yet, as recently pointed out by E. Calabrese (Clay, 1994), fewer than 200 scientists belong to both organizations! Moreover, a reading of the toxicology literature reveals a general lack of awareness within each group of the research activities of the other. Toxicologists studying fish and wildlife have much to learn from the techniques and approaches used by their colleagues in mammalian toxicology and biomedical science in general. Conversely, mammalian (rodent and human) toxicologists would benefit from a broader comparative perspective.

The objective of comparative biology is to identify the fundamental patterns and processes of life. This objective encompasses all of the various levels of

*Interconnections Between Human and Ecosystem Health*. Edited by Richard T. Di Giulio and Emily Monosson. Published in 1996 by Chapman & Hall, London. ISBN 0 412 62400 1.

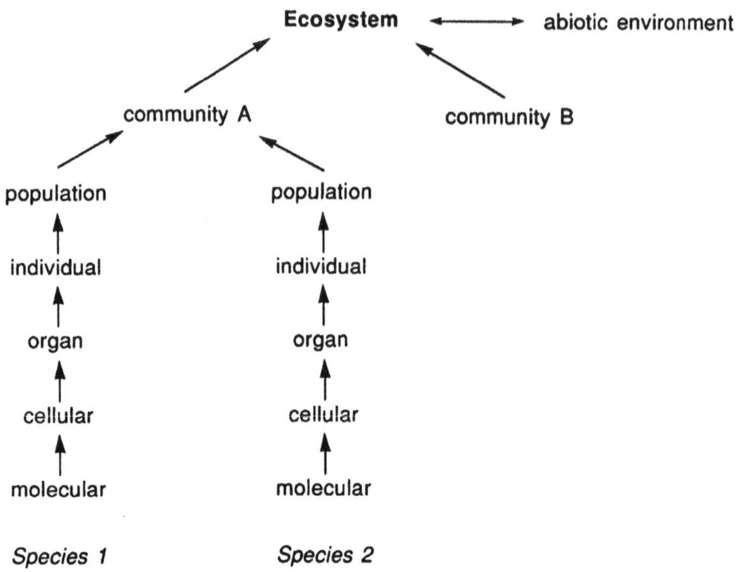


**Figure 2.1** Levels of biological organization. Environmental contaminants interact with cellular macromolecules, producing effects that are transferred through successive levels of biological organization. Effects manifest in individuals and populations of one species influence other species through altered functioning of biological communities and ecosystems.

biological organization, from the molecular/biochemical reactions involving proteins, nucleic acids, and other cellular macromolecules to the ecosystem, a complex web of interactions between populations of diverse organisms and their physical/chemical environment (Figure 2.1). Most biochemical pathways are broadly similar among different animal groups, reflecting their common evolutionary origin. Yet superimposed upon this similarity may be differences in important details. Moreover, novel pathways serving specialized functions occur in many animal groups. The challenge of comparative biochemistry is to understand both the unity and the diversity in the presence and function of these various pathways.

A comparative approach to understanding the interactions between human and ecosystem health involves the examination of the mechanisms of chemical toxicity in model experimental systems that utilize phylogenetically diverse organisms. In the past, studies carried out in such systems have led to the discovery and mechanistic description of fundamental biological principles and processes (e.g. Dawid and Sargent, 1988; Rubin, 1988; Marine Life Resources Workshop, 1989; Powers, 1989). There are multiple benefits to pursuing toxicological research on non-traditional (i.e. non-mammalian) species:

1. The information gained may be needed to determine the sensitivity of a particular group of animals to contaminants present in their environment.
2. Certain species may have unique characteristics that can be exploited to answer specific mechanistic questions. Such characteristics are only revealed, however, by research of a comparative nature.
3. From studies performed in diverse animal species, a broad phylogenetic perspective on mechanisms of toxicity can be developed, highlighting conserved features of fundamental importance and enhancing the basis for extrapolation between species.

It may be instructive to consider the parallels between assessing the risks of contaminant exposure to human health and assessing risks to the health of wildlife (as important components of ecosystems). As with humans, legal and ethical concerns may preclude the direct testing of toxic chemicals on certain types of wildlife, such as marine mammals. Therefore, alternative approaches, including extrapolation from studies conducted in other species, must be used. The process of extrapolation is most accurate when a fundamental understanding of biochemical and molecular mechanisms of action is combined with specific information on the properties of the biochemical systems in the species of interest. The latter information can often be obtained from *in vitro* or biochemical experiments that do not require the use of whole animals (e.g. White *et al.*, 1994). Thus, one of the strongest arguments for comparative, mechanistic studies is their value in providing a basis for this kind of extrapolation.

My objective in this chapter is to illustrate the contribution of a comparative toxicological approach to our understanding of the interconnections between human health and the health of other members of ecosystems. Using 'dioxin' as an example, I will show how studies carried out in a variety of animal taxa can enhance our understanding of fundamental toxic mechanisms, and in so doing, are relevant to both human and ecosystem health.

## 2.2 DIOXIN TOXICOLOGY

2,3,7,8-Tetrachlorodibenzo-*p*-dioxin (TCDD) is representative of a group of structurally related compounds – the halogenated aromatic hydrocarbons (HAH; Figure 2.2) – that are known to be highly toxic, producing a characteristic set of effects in animals exposed to them. In addition to TCDD, the HAH include other chlorinated dioxins, chlorinated dibenzofurans, chlorinated and brominated biphenyls, and several other types of halogenated aromatic compounds. Not all HAH exhibit the characteristic toxicity; there are specific structural features that determine whether specific HAH are toxic, and their potency. These features include a planar conformation as well as lateral halogen substitution (e.g. positions 2,3,7,8 of dioxins and furans; positions 3,4,5,3′,4′,5′ of biphenyls). The word 'dioxin' or 'dioxin-like' is sometimes used

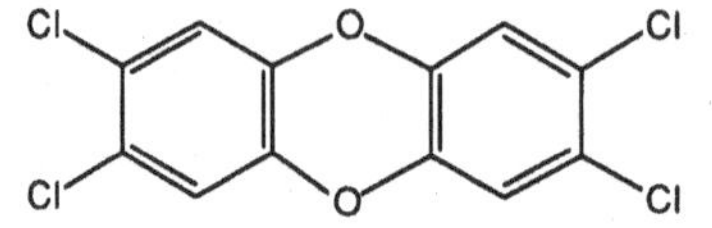

2,3,7,8-tetrachlorodibenzo-p-dioxin

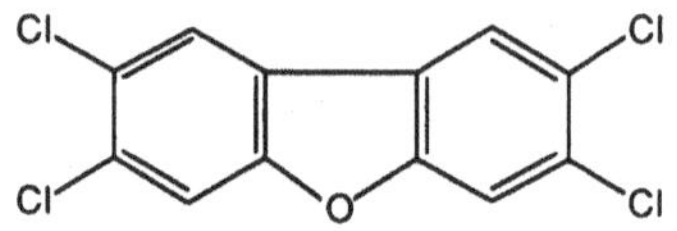

2,3,7,8-tetrachlorodibenzofuran

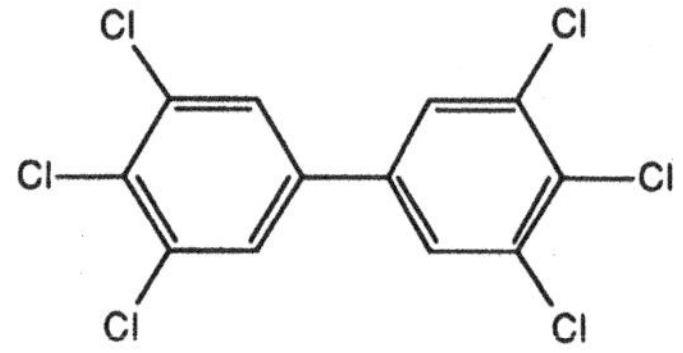

3,3',4,4',5,5'-hexachlorobiphenyl

**Figure 2.2** Structures of some representative planar halogenated aromatic hydrocarbons (PHAH).

to refer to those planar HAH (PHAH) that resemble TCDD in their effects and mode of action.

Exposure of vertebrate animals to PHAH causes a number of toxic responses including wasting, edema, thymic atrophy, other immunological effects, skin lesions (hyperkeratosis), reproductive and developmental abnormalities, endocrine dysfunction, enzyme induction and porphyria. A description of these and other effects produced by dioxin-like PHAH can be found in several excellent reviews (Poland and Knutson, 1982; Cooper, 1989; Safe, 1990; Peterson *et al.*, 1993; Pohjanvirta and Tuomisto, 1994). What is important for this chapter is not the specific nature of the dioxin effects, but how they occur and in which organisms. Of course, years of study have provided a wealth of knowledge about the response of rats, mice and other laboratory mammals to dioxins. There is an extraordinary range of sensitivities to TCDD lethality among mammalian species, with guinea pigs the most sensitive ($LD_{50}$ = 1 μg $kg^{-1}$) and hamsters the most resistant ($LD_{50}$ = 5000 μg $kg^{-1}$) (Poland and Knutson, 1982). However, the range of sensitivities of various mammalian species to the *sublethal* effects of TCDD is not nearly so broad (e.g. Gasiewicz *et al.*, 1986; Couture *et al.*, 1990; Pohjanvirta and Tuomisto, 1994), and all mammals can be considered to be 'susceptible' to dioxin effects.

Among the non-mammalian animal groups that have been examined in a laboratory setting, birds and teleost fish are also known to be highly sensitive to dioxins. In fact, the earliest indications of the extreme toxic potency of dioxins came from investigations of 'chick edema disease', an outbreak of poisoning caused by dioxin contamination of chicken feed (Firestone, 1973).

Moreover, some of the first mechanistic studies of dioxin action were carried out in chick embryos (Poland and Glover, 1973; Poland and Glover, 1977), an example of early contributions made by research in non-traditional species. The effects and potencies of dioxins in fish and birds closely resemble those seen in mammals (Sawyer *et al.*, 1986; Spitsbergen *et al.*, 1991; Walker *et al.*, 1991). Much less is known about dioxin effects in other groups of organisms, including reptiles, amphibians, 'primitive' fish such as sharks, and the many invertebrate taxa. The limited data that exist suggest that some invertebrate species are less sensitive to dioxin toxicity as compared to vertebrates (Adams *et al.*, 1986; Cooper, 1989; Dillon *et al.*, 1990; West *et al.*, 1994). However, the possibility that dioxins or other PHAH are involved in the pathogenesis of specific diseases in certain invertebrate groups is an important and ongoing area of research (see Van Beneden, Chapter 3 below).

With regard to the dioxin sensitivity of humans, there is a lively debate due to the many unanswered questions (Kimbrough, 1990; Roberts, 1991). Exposure of humans to TCDD has been well documented, but with the exception of the skin lesion chloracne, definitive links between TCDD exposure and specific effects have not been made. Studies showing expression of the Ah receptor in human cells and tissues (Manchester *et al.*, 1987; Dolwick *et al.*, 1993a) imply that the mechanisms described in rodents (see below) are also operative in humans, but the qualitative and quantitative implications (i.e. nature of effects and sensitivity, respectively) are not yet known.

As is the case with humans, the impact of dioxins on wildlife is uncertain. For some groups of animals, such as colonial, piscivorous birds in the Great Lakes and elsewhere, field and laboratory studies have provided strong evidence of PHAH effects on individuals and populations of certain species (e.g. Giesy *et al.*, 1994). For other groups, links between PHAH presence and observed disease are more tenuous (e.g. Addison, 1989; Bishop *et al.*, 1991). For example, environmental contaminants – especially PHAH – have been suggested as causative factors in several recent, well-publicized episodes of marine mammal mortality, and high levels of HAH have been reported in some cetaceans (whales) and pinnipeds (seals), but it has been extremely difficult to separate possible effects of these compounds from effects of other environmental or biological factors such as viruses or natural toxins (Dickson, 1988; Addison, 1989; Geraci *et al.*, 1989). As mentioned earlier, one way to remedy this problem is through a combination of mechanistic and comparative approaches.

## 2.3 MECHANISM OF DIOXIN ACTION: THE AH RECEPTOR

Planar HAH are thought to produce toxicity in vertebrates through changes in the expression of genes involved in the control of cell growth and differentiation. These changes in gene expression are initiated by the binding of PHAH to the aromatic hydrocarbon receptor (Ah receptor or AhR), a ligand-dependent transcription factor that is found in most vertebrate species (Swanson and Bradfield, 1993). Several of the genes whose expression is altered by PHAH have been identified (Choi *et al.*, 1991; Sutter *et al.*, 1991), but direct links between these genes and specific toxic endpoints have yet to be made. The most well-characterized gene controlled by the Ah receptor is the gene encoding cytochrome P4501A1 (CYP1A1), a member of the cytochrome P450 superfamily of heme-containing monooxygenases (Okey, 1990). Although details may differ, the model of PHAH action that has been developed by studying CYP1A1 is thought to be generally applicable to other genes controlled directly through the Ah receptor.

Three lines of evidence support the essential role of the AhR in PHAH action (Poland and Knutson, 1982; Nebert, 1989).

First, extensive studies of the structure-activity relationships for AhR binding, CYP1A1 induction and toxicity show striking parallels consistent with the idea that binding to the AhR is the initial event in PHAH action.

Second, experiments conducted in inbred strains of mice that differ at the *AHR* locus (which encodes the AhR) indicate that differential sensitivity to dioxin effects results from the expression of different allelic forms of the AhR. These AhR forms were originally distinguished by their biochemical properties, such as affinity for TCDD (Okey *et al.*, 1989; Harper *et al.*, 1991). More recently, the specific amino acid changes responsible for these biochemical differences have been identified (Chang *et al.*, 1993; Poland *et al.*, 1994).

Third, research at the molecular level has elucidated the key events involved in PHAH- and AhR-dependent gene regulation (Whitlock, 1993). This molecular understanding has been elucidated using *CYP1A1* as a model system. The sequence of events leading to induction of the *CYP1A1* gene includes binding of inducer to the AhR, transformation of the ligand-receptor complex to a DNA-binding form, interaction of the transformed ligand-receptor complex with dioxin-responsive enhancer (DRE) sequences, and transcriptional activation of target genes (Figure 2.3) (reviewed in Landers and Bunce, 1991; Whitlock, 1993). The unliganded AhR exists as a complex that includes two molecules of a 90 kDa heat-shock protein (hsp90) (Chen and Perdew, 1994). Binding of ligand to the AhR leads to dissociation of hsp90, which facilitates the time- and temperature-dependent transformation of the receptor to a form exhibiting increased affinity for specific DNA sequences. Transformation to the DNA-binding form appears to require formation of a heterodimer between the liganded AhR and another protein, the Ah receptor nuclear translocator (ARNT, Hoffman *et al.*, 1991). The transformed ligand-AhR-ARNT complex interacts with one or more DRE sequences flanking the *CYP1A1* promoter region, potentially facilitating the binding of other transcription factors and thus activating transcription of the *CYP1A1* gene.

The evidence supporting an important role for AhR-mediated events in the toxicity of PHAH suggests that the expression and function of the AhR and other proteins in the AhR signal transduction pathway could be important

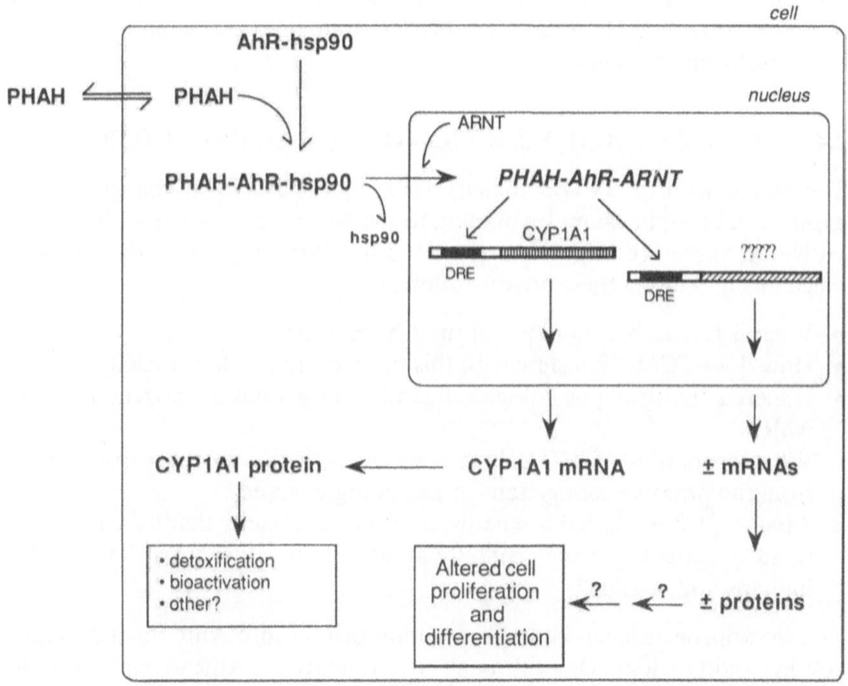


**Figure 2.3** Molecular mechanism of dioxin action mediated by the Ah receptor. Planar halogenated aromatic hydrocarbons (PHAH) bind to the Ah receptor (AhR), which is subsequently transformed to its transcriptionally active form (*PHAH-AhR-ARNT*). Transformation involves the dissociation of two molecules of the 90 kDa heat-shock protein (hsp90) and formation of a heterodimer between the AhR and the Ah receptor nuclear translocator (ARNT) protein. Binding of *PHAH-AhR-ARNT* to dioxin responsive enhancer (DRE) sequences near the promoter region of specific genes such as the gene for cytochrome P4501A1 (*CYP1A1*) initiates the transcription of those genes. Changes in gene expression can lead to toxicity through altered CYP1A1 expression or other pathways involved in the regulation of cell proliferation and differentiation. The details of dioxin effects on the latter pathways have yet to be described.

determinants of susceptibility to PHAH effects. Thus, the examination of these proteins in multiple species may help us to understand individual and species differences in dioxin sensitivity. Moreover, the information gained by studying the Ah receptor in phylogenetically diverse species is important for establishing the fundamental properties of TCDD's mechanism of action. The identification of conserved features of Ah receptor structure and function, accrued from comparative studies in diverse groups of animals, would provide increased confidence in the extrapolation of results from experimental animals to humans or to other animals (e.g. wildlife). Knowledge of Ah receptor characteristics in fish or birds, for example, is important for validating the use of such non-traditional species in toxicology. A phylogenetic approach may also enhance our understanding of the fundamental significance of the Ah receptor, providing clues to its original physiologic function and the identity of its 'endogenous' ligand.

## 2.4 PHYSIOLOGICAL FUNCTION OF THE AH RECEPTOR

The mechanism of TCDD toxicity is thought to involve changes in gene expression brought about by binding to the Ah receptor, as described above. However, there are important questions that exist regarding this proposed mechanism. Among these are the following:

- What is the normal function of the Ah receptor?
- How does TCDD interfere with this function to produce toxicity?
- Is there a 'natural' physiological ligand (endogenous or exogenous) for the AhR?
- What properties of TCDD's interaction with this receptor distinguish it from the putative endogenous or physiologic ligand?
- Are receptor-mediated mechanisms of dioxin toxicity that have been identified in rodent systems generally applicable to other organisms, such as humans and wildlife?

Two approaches can be used to obtain insight into AhR function and its role in dioxin toxicity. One approach is to compare the AhR to related proteins whose functions are known (homology approach). The other approach is to define the distribution of the AhR gene and gene products in diverse groups of organisms; similarities and differences in the physiology and cell biology of those organisms might then provide clues to AhR function (phylogenetic approach).

### *2.4.1 INSIGHTS FROM HOMOLOGY*

The function of the AhR might be elucidated in part by comparing this protein to homologous proteins (or proteins containing similar sequence motifs) whose functions are better understood. cDNAs coding for mammalian AhR and ARNT proteins have recently been cloned, revealing putative functional domains that may be involved in the protein–protein and protein–DNA interactions that are important for transcriptional control of target gene expression (Hoffman *et al.*, 1991; Burbach *et al.*, 1992; Ema *et al.*, 1992). These domains include a basic region/helix-loop-helix motif (bHLH) that is found in a large family of transcriptional regulatory proteins (Murre *et al.*, 1989) and a region known as PAS, which is a motif unique to AhR, ARNT, and the *Drosophila* proteins Per and Sim (Hoffman *et al.*, 1991; Nambu *et al.*, 1991; Burbach *et al.*, 1992) (Figure 2.4*a*,*b*).

The discovery of a bHLH motif in the AhR (Burbach *et al.*, 1992; Ema *et al.*, 1992) evoked great interest because of the large number of proteins known

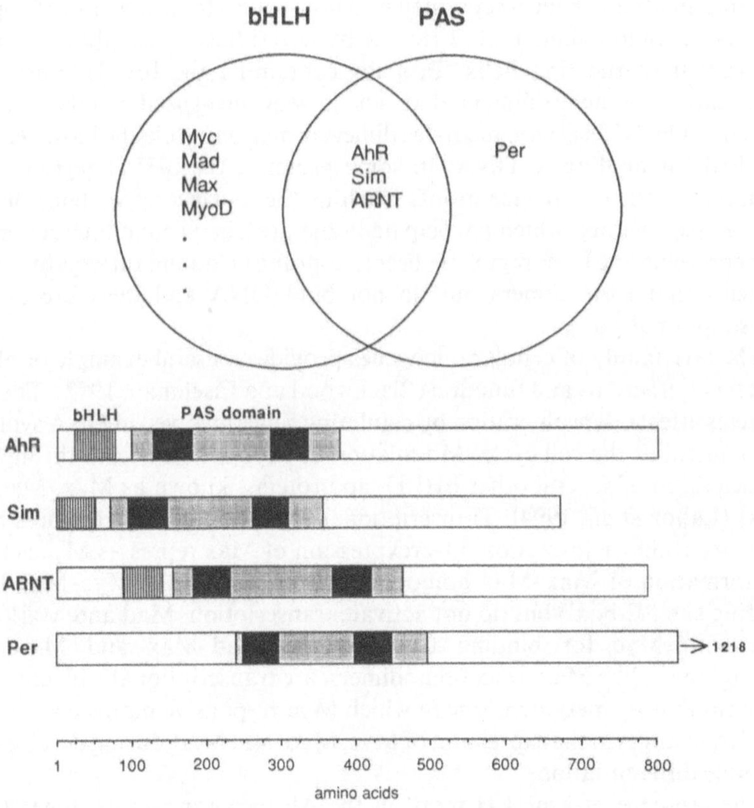


**Figure 2.4** (*Top*) Relationship of the Ah receptor to other proteins containing bHLH and PAS domains. Cloning and sequencing of the AhR (Burbach *et al.*, 1992; Ema *et al.*, 1992) has revealed its relationship to two families of proteins: the basic–helix–loop–helix (bHLH) proteins and the PAS proteins. There are more than 40 known bHLH proteins; only seven are listed here. The AhR is one of only three known proteins possessing both bHLH and PAS domains. (*Bottom*) Diagram of the PAS proteins indicating the positions of bHLH and PAS domains. The bHLH region is indicated by the vertical stripes. The stippled boxes represent the PAS domain, with the 51-amino acid repeats indicated by the black boxes. Locations of the domains are based on the alignments shown in Figures 7 and 8 of Burbach *et al.* (1992). Other functional regions are not shown.

to possess this motif and the growing understanding of their function as modulators of cell proliferation and differentiation. The bHLH protein super-family contains numerous members that are involved in a variety of functions in vertebrates and invertebrates, including neurogenesis, myogenesis, B-cell differentiation, and sex determination (reviewed in Edmondson and Olson, 1993; Kadesch, 1993).

The bHLH motif was first described by Murre *et al.* (1989). It consists of two amphipathic α-helices separated by a non-helical loop; a region of approximately 15-amino acids that is rich in basic residues is usually present just N-terminal to the first helix. Proteins containing the bHLH motif form homodimers or heterodimers that act as sequence-specific, DNA-binding proteins. The HLH region mediates dimer formation, while the basic region is required for binding to DNA. In some proteins, the bHLH region occurs adjacent to other sequence motifs, such as the leucine zipper (zip) or PAS domain (see below), which participate in the protein–protein interactions. In other proteins, the basic region is absent, resulting in 'dominant negative' HLH proteins that form dimers but do not bind DNA and therefore are not transcriptionally active.

The *myc* family of cellular oncogenes provides a useful example of bHLH protein interactions and function (Blackwood and Eisenman, 1992). The Myc proteins affect cell proliferation by regulating genes involved in DNA synthesis and control of the cell cycle. Modulation of Myc action occurs through the participation of several other bHLH-zip proteins, known as Max, Mad and Mxi1 (Lahoz *et al.*, 1994). Transcriptional activation by Myc requires Myc–Max heterodimer formation. Overexpression of Max represses Myc activity, via formation of Max–Max homodimers that bind to the Myc–Max DNA binding site ('E-box') but do not activate transcription. Mad and Mxi1 compete with Myc for binding to Max; the Mad–Max and Mxi1–Max heterodimers, like Max–Max homodimers, are transcriptionally inactive. The scenario that is emerging is one in which Myc-responsive pathways are regulated by changes in the expression of Myc, Mad, and Mxi1 during development or tissue differentiation.

The presence of a bHLH motif in the Ah receptor and its dimerization partner ARNT is consistent with the known role of this motif in protein–protein and protein–DNA interactions and the abundant evidence that the AhR–ARNT heterodimer is a sequence-specific DNA-binding complex (Whitlock, 1993). Interestingly, the consensus sequence for the dioxin-responsive enhancer (TNGCGTG, Yao and Denison, 1992) shares some similarity to the 'E-box' Myc–Max recognition sequence (CACGTG, Kadesch, 1993). The involvement of members of the bHLH superfamily in the control of cell proliferation and differentiation suggests that AhR/ARNT may also play important roles in these processes. The objective of future work will be to define those roles and to determine whether AhR and ARNT functions are modulated by other bHLH proteins, as occurs with the Myc family.

As interesting as the bHLH motif, and perhaps more significant from an evolutionary standpoint, is the presence of the PAS domain in the AhR. This domain is a 200–300 amino acid region containing two 51-amino acid, imperfect 'repeats' separated by a variable spacer region; it is found only in *P*er, *A*RNT, *A*hR and *S*im (thus, 'PAS' or 'PAAS') (Hoffman *et al.*, 1991; Nambu *et al.*, 1991; Burbach *et al.*, 1992) (Figure 2.4*b*).

The *period* (*per*) gene encodes a *Drosophila* 'clock' protein involved in maintenance of circadian rhythms (Citri *et al.*, 1987; Takahashi, 1992). Per is a nuclear protein of approximately 1200 amino acids that is unique among the four PAS proteins in lacking a bHLH region (Figure 2.4*a,b*). The *single-minded* (*sim*) gene product has an important role in the development of the *Drosophila* nervous system. The 673-amino acid Sim protein is expressed in cells along the midline of the developing central nervous system, where it controls the expression of several genes involved in CNS differentiation (Crews *et al.*, 1988; Nambu *et al.*, 1990; Nambu *et al.*, 1991). As mentioned earlier, the Ah receptor nuclear translocator (ARNT) is a mammalian protein that is required for AhR function (Hoffman *et al.*, 1991; Reyes *et al.*, 1992). Huang *et al.* (1993, 1995) showed that the PAS region of the Sim and Per proteins encompassed a novel dimerization motif that could mediate Per–Per, Per–Sim, and Per–ARNT interactions, as well as intramolecular interactions between different regions of Per. Recent studies have indicated that the PAS domain of the AhR is extremely important, not only for formation of AhR–ARNT heterodimers, but also for the binding of AhR ligands (TCDD) and interaction with hsp90 (Burbach *et al.*, 1992; Dolwick *et al.*, 1993b; Whitelaw *et al.*, 1993; Poland *et al.*, 1994; Reisz-Porszasz *et al.*, 1994).

A possible relationship between the vertebrate Ah receptor and the *Drosophila* protein Sim is most intriguing. Surprisingly, the bHLH region of the murine AhR is more closely related to that of Sim than to any other bHLH–protein, including ARNT (Ema *et al.*, 1992). The same is true when comparing the combined bHLH–PAS regions of all four PAS proteins: 31% of the residues are identical between AhR and Sim, compared to 19% and 15% for AhR versus ARNT and Per, respectively (Burbach *et al.*, 1992). Furthermore, Schmidt *et al.* (1993) found that the structures of the murine *AHR* and Drosophila *sim* genes were very similar, with four intron-exon splice junctions occurring at identical nucleotides and two others occurring close to each other. In contrast, the splice junctions in the *per* gene were very different from those of *AHR* and *sim*. These results suggest that the AhR and Sim proteins may have evolved from a common ancestor.

### 2.4.2 *INSIGHTS FROM PHYLOGENY*

Details of the AhR pathway have been learned largely through studies in mammalian systems. However, the available evidence strongly supports the idea that a similar mechanism operates in most vertebrates, including fish (reviewed in Hahn and Stegeman, 1992; Stegeman and Hahn, 1994). This evidence includes findings that:

1. Fish are among the most sensitive animals to PHAH toxicity, with lesions resembling those seen in mammals (Kleeman *et al.*, 1988; Spitsbergen *et al.*, 1991; Walker *et al.*, 1991).
2. The structure-activity relationships for HAH action in fish and birds are

similar (though not identical) to the mammalian SAR (Gooch *et al.*, 1989; Wisk and Cooper, 1990; Walker and Peterson, 1991; Hahn, 1994).
3. Animals in several vertebrate classes express Ah receptors with properties similar to those of mammalian Ah receptors (see below).

There is an extensive literature describing the Ah receptor in mammals (reviewed in Landers and Bunce, 1991; Swanson and Bradfield, 1993) but relatively little is known about the presence and properties of the Ah receptor in non-mammalian species. The Ah receptor has been identified in livers of several non-mammalian vertebrates, including chick embryos (Denison *et al.*, 1986; Poland and Glover, 1987; Brunstrom and Lund, 1988), a reptile (Hahn *et al.*, 1994), and an amphibian (Marty *et al.*, 1989). The AhR has also been found in teleost fish (Heilmann *et al.*, 1988; Lorenzen and Okey, 1990; Swanson and Perdew, 1991; Hahn *et al.*, 1992, 1993, 1994) and elasmobranch fish (Hahn *et al.*, 1994). In studies using the photoaffinity ligand 2-azido-3-[$^{125}$I]iodo-7,8-dibromodibenzo-*p*-dioxin, the size of the labeled AhR proteins in fish was shown to range from 110 to 145 kDa, similar to that seen in mammals (95–130 kDa) (Poland and Glover, 1987; Swanson and Perdew, 1991; Hahn *et al.*, 1994). Although these proteins have not been characterized in any detail, it is likely that they are structural and functional homologs of the mammalian Ah receptor.

In contrast to the results seen in teleost and elasmobranch fish, photoaffinity labeling studies have failed to reveal any specifically labeled proteins (Ah receptors) in cytosol from agnathan (jawless) fish [Atlantic hagfish (*Myxine glutinosa*) and sea lamprey (*Petromyzon marinus*)] (Hahn *et al.*, 1994). A number of invertebrate animals have also been examined for the presence of the AhR. In one study, there was no evidence of specific binding in the 50–145 kDa size range in any of several invertebrates, including three species of molluscs, two annelids, three arthropods and an echinoderm (Hahn *et al.*, 1992, 1994). Recently, specific binding of 2-azido3-[$^{125}$I]iodo-7,8-dibromodibenzo-*p*-dioxin to two low molecular weight proteins has been detected in gonads of the hard shell clam *Mercenaria* (Brown *et al.*, 1995), suggesting the existence of AhR-like proteins of unknown function in this bivalve mollusc. Some invertebrates might be expected to possess protein(s), possibly related to the *Drosophila* proteins Per and Sim (for example), that share a common ancestor with the vertebrate Ah receptor. Whether these invertebrate proteins function in any way like the vertebrate AhR is uncertain, but an intriguing possibility.

An understanding of the phylogenetic distribution of AhR proteins (Figure 2.5) may provide clues to the evolutionary origin of the *AHR* gene. The apparent absence of this protein in many marine invertebrates and in hagfish and lamprey suggests that the Ah receptor as known in vertebrates (i.e. a 95–145 kDa protein with high affinity for TCDD) may have evolved subsequent to the appearance of jawless fishes. The presence of an Ah receptor in

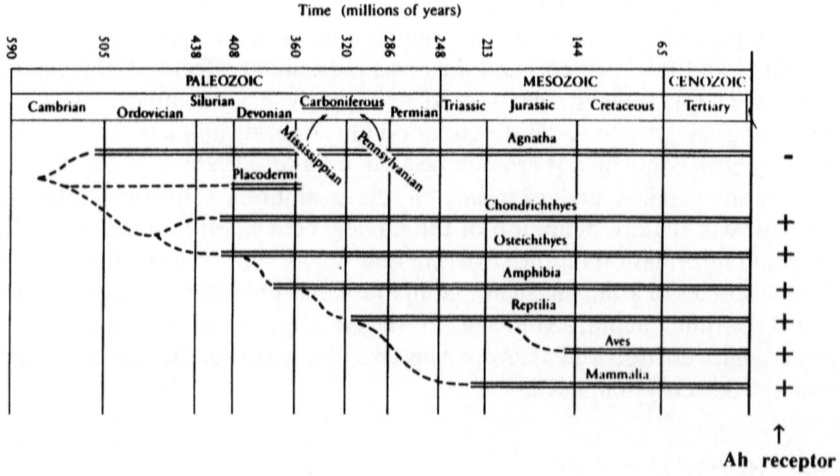


**Figure 2.5** Phylogenetic distribution of the Ah receptor in vertebrate animals. The presence (+) or absence (–) of Ah receptor proteins detectable by ligand-binding or photoaffinity labeling assays is indicated. The temporal distribution of the vertebrate classes and their probable relationships is reproduced from *Vertebrate Paleontology and Evolution* by R. L. Carroll. (Copyright © 1988 by W. H. Freeman and Company. Used with permission.) Data on AhR presence or absence are from references cited in the text. The phylogenetic distribution of the *AHR* gene is presently unknown beyond its presence in three mammals and one teleost fish.

both bony and cartilaginous fish points to an origin early in vertebrate evolution, prior to the divergence of these two groups, 450–550 million years ago (Figure 2.5) (Carroll, 1988; Forey and Janvier, 1993). Recently, an AhR cDNA from the teleost *Fundulus heteroclitus* has been partially sequenced (Hahn and Karchner, 1995). Comparison of the fish and mammalian sequences has revealed that the fish AhR is indeed homologous to mammalian Ah receptors, and that there is extensive conservation of amino acid sequences in certain regions of the AhR protein. Overall, the PAS region of the *Fundulus* AhR shares 60–65% amino acid identity with AhR proteins from mammals (Hahn and Karchner, 1995). These results suggest the existence of a strong selective pressure to maintain AhR structure and function over hundreds of millions of years.

We can only speculate on the original function of the Ah receptor and the nature of selective pressures that led to its evolution. These may have involved endogenous signaling pathways (Nebert, 1991) and/or the presence and actions of marine natural products (Stegeman and Hahn, 1994). The AhR may have evolved in concert with CYP1A1 function in a detoxification context, or alternatively, as part of a separate signal transduction system, which was only later appropriated for CYP1A1 regulation. Whether the AhR has a role in processes or pathways that help define the vertebrate condition is another intriguing question. Whatever the original function of the AhR, its similarity to bHLH and PAS proteins and the observed cellular effects of AhR ligands such as TCDD suggest that the AhR pathway has an important role in regulating cell growth and differentiation in vertebrate animals. This role has recently been confirmed (Fernandez-Salguerro *et al.*, 1995).

A more detailed understanding of the evolutionary history of the Ah receptor will require definition of the phylogenetic distribution of the *AHR* gene and information on the structure and function of its protein product in diverse species of animals. Such a comparative or phylogenetic approach will reveal the fundamental significance of the Ah receptor, providing clues to its physiologic function and a new perspective on the role of dioxins in affecting human and ecosystem health.

### *NOTE ADDED IN PROOF*

A new mammalian (human) bHLH-PAS protein, hypoxia-inducible factor-1α (HIF-1α), has recently been reported (Wang *et al.*, 1995) This protein, which is most similar to Sim, forms heterodimers with ARNT. The HIF-1α/ARNT complex appears to regulate transcription of genes, such as erythropoietin, that are involved in adaptive responses to hypoxic conditions (Wang *et al.*, 1995). The discovery of HIF-1α and its interaction with ARNT further supports the concept of multiple, interacting, bHLH-PAS proteins involved in essential cellular processes.

### *ACKNOWLEDGMENTS*

Supported in part by US/PHS grant ES06272 from the National Institute of Environmental Health Sciences (NIEHS). I would like to thank Drs Adria Elskus, Ellen Henry and Emily Monosson for their comments on the manuscript and Dr John Stegeman for many stimulating discussions. Contribution No. 8913 from the Woods Hole Oceanographic Institution.

## REFERENCES

Adams, W. J., DeGraeve, G. M., Sabourin, T. D. *et al.* (1986) Toxicity and bioconcentration of 2,3,7,8-TCDD to fathead minnows (*Pimephales promelas*). *Chemosphere*, **15**, 1503–11.

Addison, R. F. (1989) Organochlorines and marine mammal reproduction. *Canadian Journal of Fisheries and Aquatic Science*, **46**, 360–8.

Bishop, C. A., Brooks, R. J., Carey, J. H., *et al.* (1991) The case for a cause-effect linkage between environmental contamination and development in eggs of the common snapping turtle (*Chelydra s. serpentina*) from Ontario, Canada. *Journal of Toxicology and Environmental Health*, **33**, 521–47.

Blackwood, E. M. and Eisenman, R. N. (1992) Regulation of Myc:Max complex formation and its potential role in cell proliferation. *Tohoku Journal of Experimental Medicine*, **168**, 195–202.

Brown, D. J., Van Beneden, R. J. and Clark, G. C. (1995) Identification of two binding proteins for halogenated aromatic hydrocarbons in the hard-shell clam, *Mercenaria mercenaria. Archives of Biochemistry and Biophysics*, **319**, 217–24.

Brunstrom, B. and Lund, J. (1988) Differences between chick and turkey embryos in sensitivity to 3,3′,4,4′-tetrachlorobiphenyl and in concentration/affinity of the hepatic receptor for 2,3,7,8-tetrachlorodibenzo-p-dioxin. *Comparative Biochemistry and Physiology*, **91C**, 507–12.

Burbach, K. M., Poland, A. and Bradfield, C. A. (1992) Cloning of the Ah receptor cDNA reveals a distinctive ligand-activated transcription factor. *Proceedings of the National Academy of Sciences USA*, **89**, 8185–9.

Carroll, R. L. (1988) *Vertebrate Paleontology and Evolution*, New York, W. H. Freeman and Company.

Chang, C. Y., Smith, D. R., Prasad, V. S. *et al.* (1993) Ten nucleotide differences, five of which cause amino acid changes, are associated with the Ah receptor locus polymorphism of C57BL/6 and DBA/2 mice. *Pharmacogenetics*, **3**, 312–21.

Chen, H. -S. and Perdew, G. H. (1994) Subunit composition of the heteromeric cytosolic aryl hydrocarbon receptor complex. *Journal of Biological Chemistry*, **269**, 27554–8.

Choi, E. J., Toscano, D. G., Ryan, J. A. *et al.* (1991) Dioxin induces transforming growth factor-α in human keratinocytes. *Journal of Biological Chemistry*, **266**, 9591–7.

Citri, Y., Colot, H. V., Jacquier, A. C. *et al.* (1987) A family of unusually spliced biologically active transcripts encoded by a *Drosophila* clock gene. *Nature*, **326**, 42–7.

Clay, R. (1994) Assessing a damaged earth. *Environmental Health Perspectives*, **102**, 532–5.

Cooper, K. R. (1989) Effects of polychlorinated dibenzo-*p*-dioxins and polychlorinated dibenzofurans on aquatic organisms. *CRC Critical Reviews in Aquatic Sciences*, **1**, 227–42.

Couture, L. A., Abbott, B. D. and Birnbaum, L. S. (1990) A critical review of the developmental toxicity and teratogenicity of 2,3,7,8-tetrachlorodibenzo-p-dioxin: recent advances toward understanding the mechanism. *Teratology*, **42**, 619–27.

Crews, S. T., Thomas, J. B. and Goodman, C. S. (1988) The Drosophila *single-minded* gene encodes a nuclear protein with sequence similarity to the *per* gene product. *Cell*, **52**, 143–51.

Dawid, I. B. and Sargent, T. D. (1988) *Xenopus laevis* in developmental and molecular biology. *Science*, **240**, 1443–8.

Denison, M. S., Wilkinson, C. F. and Okey, A. B. (1986) Ah receptor for 2,3,7,8-tetrachlorodibenzo-p-dioxin: ontogeny in chick embryo liver. *Journal of Biochemical Toxicology*, **1**, 39–49.

Dickson, D. (1988) Mystery disease strikes Europe's seals. *Nature*, **241**, 893–5.

Dillon, T. M., Benson, W. H., Stackhouse, R. A. and Crider, A. M. (1990) Effects of selected PCB congeners on survival, growth, and reproduction in *Daphnia magna. Environmental Toxicology and Chemistry*, **9**, 1317–26.

Dolwick, K. M., Schmidt, J. V., Carver, L. A. *et al.* (1993a) Cloning and expression of a human Ah receptor cDNA. *Molecular Pharmacology*, **44**, 911–17.

Dolwick, K. M., Swanson, H. I. and Bradfield, C. A. (1993b) *In vitro* analysis of Ah receptor domains involved in ligand-activated DNA recognition. *Proceedings of the National Academy of Sciences USA*, **90**, 8566–70.

Edmondson, D. G. and Olson, E. N. (1993) Helix-loop-helix proteins as regulators of muscle-specific transcription. *Journal of Biological Chemistry*, **268,** 755–8.

Ema, M., Sogawa, K., Watanabe, N. *et al.* (1992) cDNA cloning and structure of mouse putative Ah receptor. *Biochemical and Biophysical Research Communications*, **184,** 246–53.

Fernandez-Salguerro, P., Pineau, T., Hilbert, D. M. *et al.* (1995) Immune system impairment and hepatic fibrosis in mice lacking the dioxin-binding Ah receptor. *Science*, **268,** 722–6.

Firestone, D. (1973) Etiology of chick edema disease. *Environmental Health Perspectives*, **5,** 59–66.

Forey, P. and Janvier, P. (1993) Agnathans and the origin of jawed vertebrates. *Nature*, **361,** 129–34.

Gasiewicz, T. A., Rucci, G., Henry, E. C. and Baggs, R. B. (1986) Changes in hamster hepatic cytochrome P450, ethoxycoumarin O-deethylase, and reduced NAD(P): menadione oxidoreductase following treatment with 2,3,7,8-tetrachlorodibenzo-p-dioxin. *Biochemical Pharmacology*, **35,** 2737–42.

Geraci, J. R., Anderson, D. M., Timperi, R. J. *et al.* (1989) Humpback whales (*Megaptera novaeangliae*) fatally poisoned by dinoflagellate toxin. *Canadian Journal of Fisheries and Aquatic Science*, **46,** 1895–8.

Giesy, J. P., Ludvig, J. P. and Tillitt, D. E. (1994) Deformities in birds of the Great Lakes Region: assigning causality. *Environmental Science and Technology*, **28,** 128–35.

Gooch, J. W., Elskus, A. A., Kloepper-Sams, P. J. *et al.* (1989) Effects of *ortho* and non-*ortho* substituted polychlorinated biphenyl congeners on the hepatic monooxygenase system in scup (*Stenotomus chrysops*). *Toxicology and Applied Pharmacology*, **98,** 422–33.

Hahn, M. E. (1994) Cytochrome P450 induction and inhibition by planar halogenated aromatic hydrocarbons in a fish cell line: promise and pitfalls for environmental testing. *In Vitro Cell and Developmental Biology*, **30A,** 39–40.

Hahn, M. E. and Karchner, S. I. (1995) Evolutionary conservation of the vertebrate Ah (dioxin) receptor: amplification and sequencing of the PAS domain of a teleost Ah receptor cDNA. *Biochemical Journal*, **310,** 383–7.

Hahn, M. E. and Stegeman, J. J. (1992) Phylogenetic distribution of the Ah receptor in non-mammalian species: implications for dioxin toxicity and Ah receptor evolution. *Chemosphere*, **25,** 931–7.

Hahn, M. E., Lamb, T. M., Schultz, M. E. *et al.* (1993) Cytochrome P4501A induction and inhibition by 3,3′,4,4′-tetrachlorobiphenyl in an Ah receptor-containing fish hepatoma cell line (PLHC-1). *Aquatic Toxicology*, **26,** 185–208.

Hahn, M. E., Poland, A., Glover, E. and Stegeman, J. J. (1992) The Ah receptor in marine animals. Phylogenetic distribution and relationship to P4501A inducibility. *Marine Environmental Research*, **34,** 87–92.

Hahn, M. E., Poland, A., Glover, E. and Stegeman, J. J. (1994) Photoaffinity labeling of the Ah receptor. Phylogenetic survey of diverse vertebrate and invertebrate species. *Archives of Biochemistry and Biophysics*, **310,** 218–28.

Harper, P. A., Golas, C. L. and Okey, A. B. (1991) Ah receptor in mice genetically 'nonresponsive' for cytochrome P4501A1 induction: cytosolic Ah receptor, transformation to the nuclear binding state, and induction of aryl hydrocarbon hydroxylase by halogenated and nonhalogenated hydrocarbons in embryonic tissues and cells. *Molecular Pharmacology*, **40,** 818–26.

Heilmann, L. J., Sheen, Y. Y., Bigelow, S. W. and Nebert, D. W. (1988) The trout

P450IA1: cDNA and deduced protein sequence, expression in liver, and evolutionary significance. *DNA*, **7**, 379–87.

Hoffman, E. C., Reyes, H., Chu, F. -F. *et al.* (1991) Cloning of a factor required for activity of the Ah (dioxin) receptor. *Science*, **252**, 954–8.

Huang, Z. J., Curtin, K. D. and Rosbash, M. (1995) PER protein interactions and temperature compensation of a circadian clock in *Drosophila*. *Science*, **267**, 1169–72.

Huang, Z. J., Edery, I. and Rosbash, M. (1993) PAS is a dimerization domain common to *Drosophila* period and several transcription factors. *Nature*, **364**, 259–62.

Kadesch, T. (1993) Consequences of heteromeric interactions among helix-loop-helix proteins. *Cell Growth and Differentiation*, **4**, 49–55.

Kimbrough, R. D. (1990) How toxic is 2,3,7,8-tetrachlorodibenzo-p-dioxin to humans? *Journal of Toxicology and Environmental Health*, **30**, 261–71.

Kleeman, J. M., Olson, J. R. and Peterson, R. E. (1988) Species differences in 2,3,7,8-tetrachlorodibenzo-*p*-dioxin toxicity and biotransformation in fish. *Fundamental and Applied Toxicology*, **10**, 206–13.

Lahoz, E. G., Xu, L., Schreiber-Agus, N. and DePinho, R. A. (1994) Suppression of Myc, but not E1a, transformation activity by Max-associated proteins, Mad and Mxi1. *Proceedings of the National Academy of Sciences USA*, **91**, 5503–7.

Landers, J. P. and Bunce, N. J. (1991) The *Ah* receptor and the mechanism of dioxin toxicity. *Biochemical Journal*, **276**, 273–87.

Lorenzen, A. and Okey, A. B. (1990) Detection and characterization of [$^3$H]2,3,7,8-tetrachlorodibenzo-*p*-dioxin binding to Ah receptor in a rainbow trout hepatoma cell line. *Toxicology and Applied Pharmacology*, **106**, 53–62.

Manchester, D. K., Gordon, S. K., Golas, C. L. *et al.* (1987) Ah receptor in human placenta: stabilization by molybdate and characterization of binding of 2,3,7,8-tetrachlorodibenzo-p-dioxin, 3-methylcholanthrene, and benzo[a]pyrene. *Cancer Research*, **47**, 4861–8.

Marine Life Resources Workshop (1989) Marine models in biomedical research. *Biological Bulletin*, **176**, 337–48.

Marty, J., Lesca, P., Jaylet, A. *et al.* (1989) In vivo and in vitro metabolism of benzo(a)pyrene by the larva of the newt, Pleurodeles waltl. *Comparative Biochemistry and Physiology*, **93C**, 213–19.

Murre, C., McCaw, P. S. and Baltimore, D. (1989) A new DNA binding and dimerization motif in immunoglobulin enhancer binding, *daughterless, MyoD*, and *myc* proteins. *Cell*, **56**, 777–83.

Nambu, J. R., Franks, R. G., Hu, S. and Crews, S. T. (1990) The *single-minded* gene of Drosophila is required for the expression of genes important for the development of CNS midline cells. *Cell*, **63**, 63–75.

Nambu, J. R., Lewis, J. O., Wharton, K. A. and Crews, S. T. (1991) The Drosophila *single-minded* gene encodes a helix-loop-helix protein that acts as a master regulator of CNS midline development. *Cell*, **67**, 1157–67.

Nebert, D. W. (1989) The Ah locus: genetic differences in toxicity, cancer, mutation, and birth defects. *CRC Critical Reviews in Toxicology*, **20**, 137–52.

Nebert, D. W. (1991) Proposed role of drug-metabolizing enzymes: regulation of steady state levels of the ligands that effect growth, homeostasis, differentiation, and neuroendocrine functions. *Molecular Endocrinology*, **5**, 1203–14.

Okey, A. B. (1990) Enzyme induction in the cytochrome P-450 system. *Pharmacology and Therapeutics*, **45**, 241–98.

Okey, A. B., Vella, L. M. and Harper, P. A. (1989) Detection and characterization of a low affinity form of cytosolic Ah receptor in livers of mice nonresponsive to induction of cytochrome $P_1$-450 by 3-methylcholanthrene. *Molecular Pharmacology*, **35**, 823–30.

Peterson, R. E., Theobald, H. M. and Kimmel, G. L. (1993) Developmental and reproductive toxicity of dioxins and related compounds – cross-species comparisons. *CRC Critical Reviews in Toxicology*, **23**, 283–335.

Pohjanvirta, R. and Tuomisto, J. (1994) Short-term toxicity of 2,3,7,8-tetrachlorodibenzo-p-dioxin in laboratory animals: effects, mechanisms, and animal models. *Pharmacological Reviews*, **46**, 483–549.

Poland, A. and Glover, E. (1973) Chlorinated dibenzo-p-dioxins: potent inducers of delta-aminolevulinic acid synthetase and aryl hydrocarbon hydroxylase. II. A study of the structure-activity relationship. *Molecular Pharmacology*, **9**, 736–47.

Poland, A. and Glover, E. (1977) Chlorinated biphenyl induction of aryl hydrocarbon hydroxylase activity: a study of the structure-activity relationship. *Molecular Pharmacology*, **13**, 924–38.

Poland, A. and Glover, E. (1987) Variation in the molecular mass of the Ah receptor among vertebrate species and strains of rats. *Biochemical and Biophysical Research Communications*, **146**, 1439–49.

Poland, A. and Knutson, J. C. (1982) 2,3,7,8-Tetrachlorodibenzo-p-dioxin and related halogenated aromatic hydrocarbons: examination of the mechanism of toxicity. *Annual Reviews of Pharmacology and Toxicology*, **22**, 517–54.

Poland, A., Palen, D. and Glover, E. (1994) Analysis of the four alleles of the murine aryl hydrocarbon receptor. *Molecular Pharmacology*, **46**, 915–21.

Powers, D. A. (1989) Fish as model systems. *Science*, **246**, 352–8.

Reisz-Porszasz, S., Probst, M. R., Fukunaga, B. N. and Hankinson, O. (1994) Identification of functional domains of the aryl hydrocarbon receptor nuclear translocator protein (ARNT). *Molecular and Cellular Biology*, **14**, 6057–86.

Reyes, H., Reisz-Porszasz, S. and Hankinson, O. (1992) Identification of the Ah receptor nuclear translocator protein (Arnt) as a component of the DNA binding form of the Ah receptor. *Science*, **256**, 1193–5.

Roberts, L. (1991) Dioxin risks revisited. *Science*, **251**, 624–6.

Rubin, G. M. (1988) *Drosophila melanogaster* as an experimental organism. *Science*, **240**, 1453–9.

Safe, S. (1990) Polychlorinated biphenyls (PCBs), dibenzo-p-dioxins (PCDDs), dibenzofurans (PCDFs), and related compounds: environmental and mechanistic considerations which support the development of toxic equivalency factors (TEFs). *CRC Critical Reviews in Toxicology*, **21**, 51–88.

Sawyer, T., Jones, D., Rosanoff, K. *et al.* (1986) The biologic and toxic effects of 2,3,7,8-tetrachlorodibenzo-p-dioxin in chickens. *Toxicology*, **39**, 197–206.

Schmidt, J. V., Carver, L. A. and Bradfield, C. A. (1993) Molecular characterization of the murine *Ahr* gene. Organization, promoter analysis, and chromosomal assignment. *Journal of Biological Chemistry*, **268**, 22203–9.

Spitsbergen, J. M., Walker, M. K., Olson, J. R. and Peterson, R. E. (1991) Pathologic alterations in early life stages of lake trout, *Salvelinus namaycush*, exposed to 2,3,7,8-tetrachlorodibenzo-*p*-dioxin as fertilized eggs. *Aquatic Toxicology*, **19**, 41–72.

Stegeman, J. J. and Hahn, M. E. (1994) Biochemistry and molecular biology of monooxygenases: current perspectives on forms, functions, and regulation of cytochrome P450 in aquatic species, in *Aquatic Toxicology: Molecular Biochemical and Cellular Perspectives*, (eds D. C. Malins and G. K. Ostrander), CRC/Lewis, Boca Raton, pp. 87–206.

Sutter, T. R., Guzman, K., Dold, K. M. and Greenlee, W. F. (1991) Targets for dioxin: genes for plasminogen activator inhibitor-2 and interleukin-1 β. *Science*, **254**, 415–18.

Swanson, H. I. and Bradfield, C. A. (1993) The AH-receptor – genetics, structure and function. *Pharmacogenetics*, **3**, 213–30.

Swanson, H. I. and Perdew, G. H. (1991) Detection of the Ah receptor in rainbow trout. Use of 2-azido-3-[$^{125}$I]iodo-7,8-dibromodibenzo-p-dioxin in cell culture. *Toxicology Letters*, **58**, 85–95.

Takahashi, J. S. (1992) Circadian clock genes are ticking. *Science*, **258**, 238–40.

Walker, M. K. and Peterson, R. E. (1991) Potencies of polychlorinated dibenzo-*p*-dioxin, dibenzofuran, and biphenyl congeners, relative to 2,3,7,8-tetrachlorodibenzo-p-dioxin, for producing early life stage mortality in rainbow trout (*Oncorhynchus mykiss*). *Aquatic Toxicology*, **21**, 219–38.

Walker, M. K., Spitsbergen, J. M., Olson, J. R. and Peterson, R. E. (1991) 2,3,7,8-Tetrachlorodibenzo-*p*-dioxin (TCDD) toxicity during early life stage development of lake trout (*Salvelinus namaycush*). *Canadian Journal of Fisheries and Aquatic Science*, **48**, 875–83.

Wang, G. L., Jiang, B. -H., Rue, E. A. and Semenza, G. L. (1995) Hypoxia-inducible factor 1 is a basic-helix-loop-helix-PAS heterodimer regulated by cellular $O_2$ tension. *Proceedings of the National Academy of Sciences USA*, **92**, 5510–14.

West, C. W., Ankley, G. T., Elonen, G. E. *et al.* (1994) Toxicity and bioaccumulation of 2,3,7,8-TCDD by *Chironomus tentans* and *Lumbriculus variegatus*. Society of Environmental Toxicology and Chemistry, 15th Annual Meeting, Denver, CO.

White, R. D., Hahn, M. E., Lockhart, W. L. and Stegeman, J. J. (1994) Catalytic and immunochemical characterization of hepatic microsomal cytochromes P450 in beluga whales (*Delphinapterus leucas*). *Toxicology and Applied Pharmacology*, **126**, 45–57.

Whitelaw, M., Gottlicher, M., Gustafsson, J. A. and Poellinger, L. (1993) Definition of a novel ligand binding domain of a nuclear bHLH receptor: co-localization of ligand and hsp90 binding activities within the regulable inactivation domain of the dioxin receptor. *EMBO Journal*, **12**, 4169–79.

Whitlock, J. P. (1993) Mechanistic aspects of dioxin action. *Chemical Research in Toxicology*, **6**, 754–63.

Wisk, J. D. and Cooper, K. R. (1990) Comparison of the toxicity of several polychlorinated dibenzo-*p*-dioxins and 2,3,7,8-tetrachlorodibenzofuran in embryos of the Japanese medaka (*Oryzias latipes*). *Chemosphere*, **20**, 361–77.

Yao, E. F. and Denison, M. S. (1992) DNA sequence determinants for binding of transformed Ah-receptor to a dioxin-responsive enhancer. *Biochemistry*, **31**, 5060–7.

# 3 Comparative studies of molecular mechanisms of tumorigenesis in herbicide-exposed bivalves

REBECCA J. VAN BENEDEN

## 3.1 INTRODUCTION

The degree to which exposure to environmental toxicants induces cancer has long been a subject of debate. In most cases, direct causality has been difficult, if not impossible, to prove. This is largely because carcinogenesis is a very complex, multistep process involving both external factors imposed by the environment as well as the intrinsic genetic background of the individual. One approach to understanding this process is to examine the basic underlying mechanisms of chemical carcinogenesis. Recent rapid technological advances in recombinant DNA analysis have had a tremendous impact on basic research in the molecular mechanisms of chemical carcinogenesis (Yupsa and Poirier, 1988). Along with advances at the molecular level has come the realization that most of these mechanisms are evolutionarily conserved. This insight has greatly expanded the number of model systems which can be used to study pathological changes in common molecular pathways of cell growth and differentiation. One of the most promising areas in the development of alternative, non-mammalian models is the use of aquatic models.

Since the 1960s, epidemiological research of aquatic animals has demonstrated a correlation between the incidence of tumors and environmental exposure to chemicals (Harshbarger and Clark, 1990; Gardner, 1993). Tumor incidence has reached alarming proportions in English sole (*Parophyrs vetulus*) in certain areas of Puget Sound (Krahn *et al.*, 1986), winter flounder (*Pseudopleuronectes americanus*) in the Boston Harbor (Murchelano and Wolke, 1985), brown bullhead catfish (*Ictalurus nebulosus*) in the Black,

*Interconnections Between Human and Ecosystem Health.* Edited by Richard T. Di Giulio and Emily Monosson. Published in 1996 by Chapman & Hall, London. ISBN 0 412 62400 1.

Niagara and Buffalo Rivers (Baumann *et al.*, 1982; Black, 1983), and Atlantic tomcod (*Microgadus tomcod*) in the Hudson River (Smith *et al.*, 1979). It is important to emphasize that although exposure to chemical carcinogens is highly correlated with tumor incidence, no direct proof of the etiology of these tumors exists.

This chapter presents the preliminary results from our laboratory and others of the investigation of the molecular mechanism of gonadal neoplasms in two marine bivalve species. These bivalves represent important model species which are in direct contact with inshore waters and sediments and thus are excellent candidates for sentinel species. While the initial descriptions of these tumors were made about 20 years ago (Yevich and Barry, 1967; Yevich and Barszcz, 1977), the etiology of these tumors is unknown. Both chemical exposure and genetic factors have been proposed and are discussed below. Studies of the molecular mechanisms of clam gonadal tumors should allow comparisons to those of higher organisms, including humans.

## 3.2 INITIAL STUDIES OF GONADAL NEOPLASMS IN BIVALVES

In the early to mid 1970s, results of field surveys by the EPA Environmental Research Laboratory in Narragansett suggested a possible association between herbicide contamination and epizootic germinomas in three geographically distinct populations of soft-shell clams (*Mya arenaria*) in eastern Maine (Figure 3.1*a*; Gardner *et al.*, 1991b): Roque Bluffs, Hardscrabble River and Hobart Stream. At Roque Bluffs, 3% of the animals had gonadal tumors. The prevalence of germinomas in soft-shell clams collected from three areas of the Hardscrabble River area near Dennysville was 32%, 32% and 40%, respectively. Seven per cent of the animals collected from this area also had supernumerary siphons. In three collections from the Hobart Stream near the Moosehorn Wildlife Refuge, tumor prevalence was 34%, 28% and 36%. Both the Hardscrabble River and Hobart Stream empty into Dennys Bay. Investigation into the etiology of these tumors revealed that all three locations in Maine were subjected to herbicide exposure. Significant quantities of Tordon 101 (picloram), 2,4-D (2,4,-dichlorophenoxyacetic acid) and 2,4,5-T (2,4,5-trichlorophenoxyacetic acid) had been used in these areas in blueberry culture, road maintenance and forestry (Gardner *et al.*, 1991b). While 2,4-D and 2,4,5-T have exhibited low toxicity in mammalian assays, TCDD (dioxin; 2,3,7,8-tetrachlorodibenzo-*p*-dioxin), a by-product contaminant from the synthesis of 2,4,5-T, has been described as one of the most toxic environmental contaminants known (Schmidt, 1992).

Histologically similar tumors have also been reported in hard-shell clams (*Mercenaria spp.*) taken from the Indian River, Florida (Figure 3.1 *b*; Hesselman *et al.*, 1988). The Indian River estuary is a poorly flushed 200 mile long estuarine area that receives drainage from central Florida agricultural areas. In some areas, tumor prevalence was greater than 60%, the highest level of neoplasia ever reported in an invertebrate. This study was unable to establish the cause of the high incidence of gonadal tumors. However, >50 herbicide and

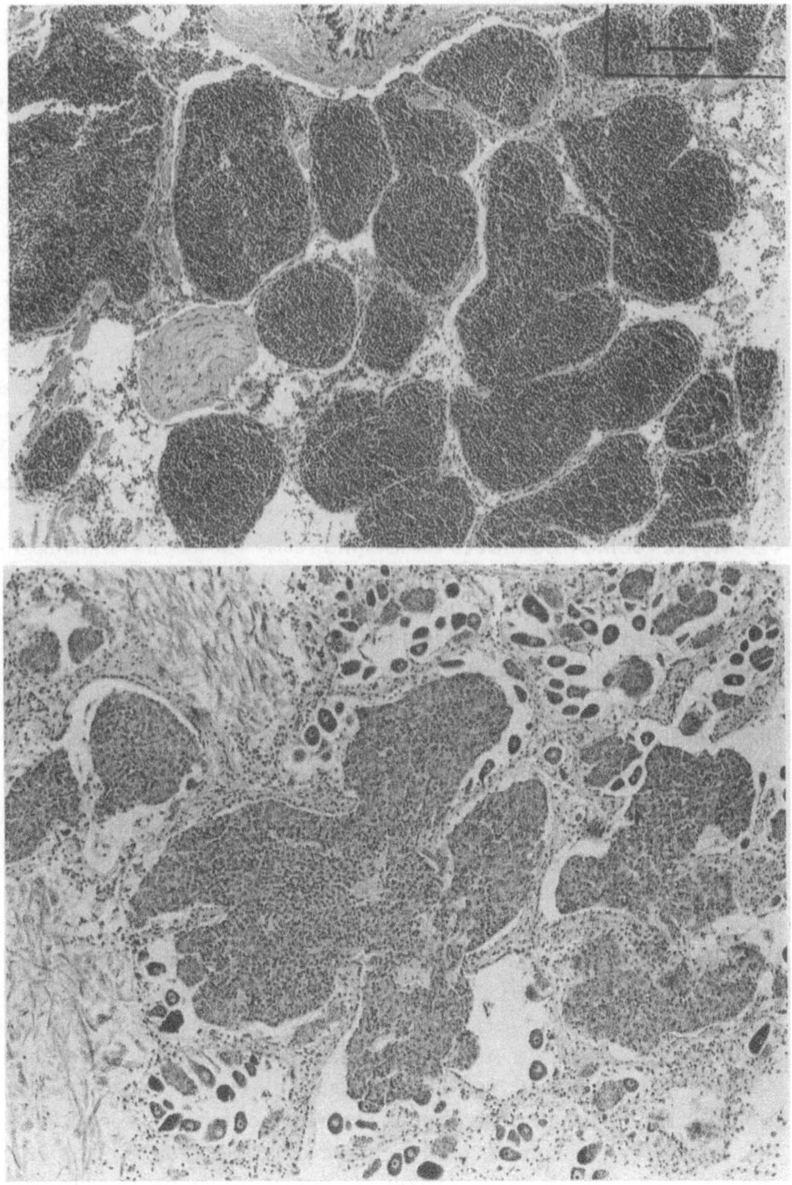

(a)

(b)

**Figure 3.1** (a) *Mya arenaria* male germinal neoplasm. The typical seminoma presents testicular follicles with the same type of uniform, undifferentiated germ cells seen in dysgerminomas. Seminomas and dysgerminomas sometimes metastasize widely. Tissue was prepared as described previously (Gardner *et al.*, 1991b) and stained with hematoxylin and eosin (100×). Scale bar = 80 μm. (b) *Mercenaria mercenaria* neoplasm of ovarian germinal tissue. Detail of ovary reveals normal oocyte association with ovarian follicle, normal oocytes and clustered neoplastic germ cells, and neoplastic cells demarcated by follicles distended by the tumor. The typical dysgerminoma is composed of a single monomorphic, basophilic germ cell type. Tissue was prepared as described previously (Gardner *et al.*, 1991b) and stained with hematoxylin and eosin (100×).

pesticide contaminants have been identified in sediments from the Indian River drainage system (Harshbarger and Clark, 1990). This estuarine system drains a wide area of citrus groves and receives substantial domestic and agricultural runoff. This suggests that is possible that the clams in the Indian River had been exposed to herbicides (see also Section 3.7 Genetic Elements).

The suggested association between herbicides and molluscan gonadal neoplasms was strengthened by critical laboratory studies of bivalves that linked the rapid uptake of chemical carcinogens and mutagenic activity to tumor formation (Pittenger *et al.*, 1985, 1987; Gardner *et al.*, 1991a). Oysters (*Crassostrea virginica*) developed a variety of neoplasms when exposed to carcinogen contaminated sediments from Black Rock Harbor, CT, both in the field and in the laboratory (Gardner *et al.*, 1991a). Both oysters and blue mussels (*Mytilus edulis*) accumulated high concentrations of polychlorinated biphenyls, polyaromatic hydrocarbons and chlorinated pesticides. Winter flounder fed blue mussels grown on contaminated sediments developed both renal and pancreatic tumors, which demonstrated that sediment-bound carcinogens could also induce tumors indirectly via trophic transfer.

Additional reports suggest that this phenomenon is not limited to poikilotherms. Seminomas have been detected in military working dogs believed to be exposed to herbicides in Vietnam (Hayes *et al.*, 1990). A second study (Hayes *et al.*, 1991) showed a positive correlation between the incidence of malignant lymphomas in dogs and their owners' use of 2,4-D containing herbicides on their lawns. In other studies, Rhesus monkeys exposed to chronic, low levels of dioxin were found to have an increased rate of endometriosis (Rier *et al.*, 1993). Exposure to TCDD was directly correlated to the incidence of endometriosis and the severity was dose-dependent. Preliminary studies of female veterans who served in the Vietnam War indicated a higher than expected mortality rate from pancreatic and uterine cancer (Matthews, 1992). Furthermore, an EPA survey of cancer mortality rates in the US indicated that the mortality rate due to ovarian cancer in human females from Washington County, ME, and near Indian River, FL, was significantly higher than the national average (Riggan *et al.*, 1987). These are the same geographical areas in which the tumor-bearing clam populations are located. The observation of increased rates of human gonadal cancers in the same areas and reproductive dysfunctions in dogs and monkeys exposed to herbicides suggests that a common mechanism might be involved in these related neoplasms. One approach to test this hypothesis would be comparative studies of the molecular mechanisms of toxicity of TCDD and other aryl hydrocarbons.

## 3.3 DIOXIN TOXICITY

TCDD and other polyhalogenated aromatic hydrocarbons are man-made compounds which are now ubiquitous, low-level environmental contaminants. The widespread occurrence is largely the result of the use of contaminated herbicides, high-temperature combustion processes and chlorine bleaching of wood products. Dioxins were first recognized as contaminants in some chemical preparations in the late 1950s. Early tests to determine the toxicity of these chemicals indicated that dioxins were among the most potent man-made carcinogens known. Although most sources of dioxin have been identified and eliminated, these are very stable compounds which can remain in sediments for many years.

Laboratory animals exposed to dioxin exhibit extreme variations in sensitivity (Table 3.1; Hanson, 1991) as well as a wide range of symptoms including decreased appetite, immunosuppression, chloracne, teratogenicity, carcinogenicity, immunotoxicity and death. Biochemical effects include activation of drug-metabolizing enzymes and growth factor pathways (Poland and Knutson, 1982; Silbergeld and Gasiweicz, 1989; Zeise *et al.*, 1990; Landers and Bunce, 1991). In spite of numerous laboratory studies, there are still few conclusive data concerning the responses of humans to dioxins. In fact, the position that humans occupy on the sensitivity scale is still unknown.

The molecular basis for the toxicity and carcinogenicity of TCDD in vertebrates has been extensively investigated. In 1976, a clue to the mechanism of these responses was provided by the discovery of the Ah receptor (aryl hydrocarbon receptor) – a high-affinity cytoplasmic receptor which bound not only dioxin but other polycyclic aromatic hydrocarbons (Poland *et al.*, 1976). Most symptoms observed in animals exposed to this class of compounds are believed to arise from their interaction with this receptor (Roberts, 1991). The current proposed mechanism of this interaction is illustrated in Figure 3.2. The inactive Ah receptor is found in the cytoplasm bound to two molecules of the stress protein HSP90. Binding of a ligand to the Ah receptor results in dissociation of the HSP90 molecules and the association of at least one other protein, arnt (Ah Receptor Nuclear Translocator). The activated receptor complex is then translocated to the nucleus where it binds to specific dioxin responsive elements (DREs or XREs, xenobiotic responsive elements) upstream of specific genes where it acts as a transcriptional activator. The first

**Table 3.1** Species variation to TCDD toxicity

| Species | $LD_{50}$ (μg/kg) |
|---|---|
| Guinea pig | 0.6–2.5 |
| Mink | 4 |
| Rat | 22–320 |
| Monkey | <70 |
| Rabbit | 115–275 |
| Mouse | 114–280 |
| Dog | >100–<3000 |
| Hamster | 1150–5000 |

TCDD – Dioxin; 2,3,7,8-tetrachlorodibenzo-p-dioxin.
*Source*: Adapted from Hanson (1991).

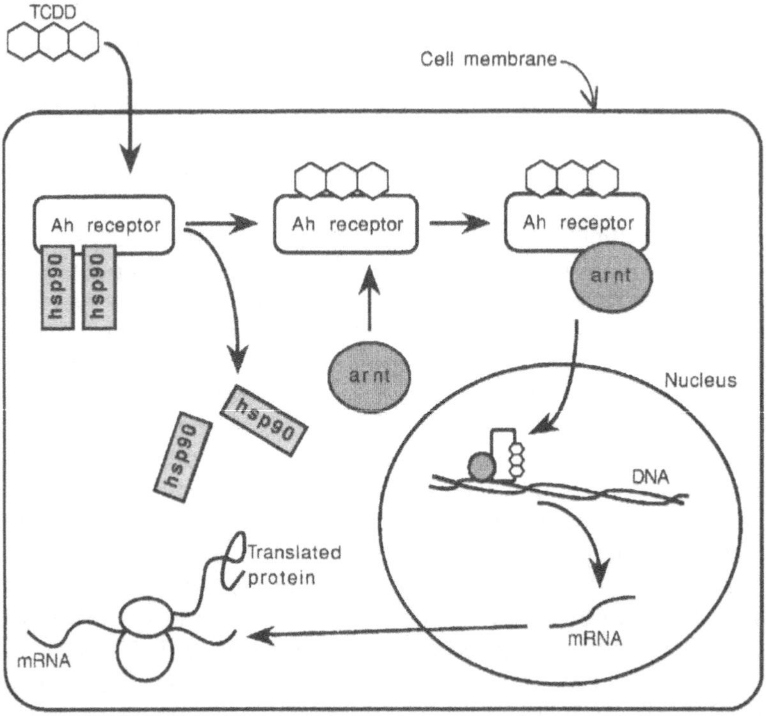


**Figure 3.2** Schematic of currently proposed mechanism of transcriptional activation by the Ah receptor/ligand complex.

described and best understood is the activation of the cytochrome P450 1A1 (Cyp1A1) gene (Whitlock, 1990; Hankinson, 1993). Alteration of transcription of other TCDD-responsive genes is hypothesized to be the mechanistic basis by which TCDD and other PAHs express their biological effects (Sutter and Greenlee, 1992). Studies of the recently cloned mouse *Ah* receptor confirm its role as a transcriptional activator (Burbach *et al.*, 1992).

## 3.4 THE SEARCH FOR THE INVERTEBRATE ANALOG OF THE AH RECEPTOR

Studies are under way in our laboratory to determine whether these bivalves have an Ah receptor-like protein and to define the role that it may play in these tumors. The Ah receptor has been identified in a number of vertebrate species including humans, rodents and trout (Lorenzen and Okey, 1990). Hahn *et al.* (1994) reported that the Ah receptor was found in a number of different species of marine fish, but was not present in the nine invertebrate species that they examined (they did not use either *Mercenaria sp.* or *Mya arenaria*). The detection of a homologous dioxin receptor system in an invertebrate would suggest conservation over a long evolutionary time period and a critical role for the Ah receptor.

Studies were initiated in our laboratory to determine whether the clams possessed an Ah receptor system analogous to that reported in vertebrates. For these studies, cytosols were prepared from *M. mercenaria* from several different tissue types, including gonadal tissue from both male and females (Brown *et al.*, 1995). Cytosol fractions were labeled with the photoaffinity analog, 2-azido-3-[$^{125}$I]-iodo-7,8-dibromodibenzo-*p*-dioxin by the method of Poland *et al.* (1986) with minor modifications as described previously (Brown *et al.*, 1995). Precipitated samples were run on SDS-polyacrylamide gels (Laemmli, 1970) and subjected to autoradiography (Figure 2.4). Photoaffinity labeling resulted in a number of radiolabeled bands, however, only two bands were blocked by the presence of competing ligand. The molecular weights of these two proteins were determined by SDS-PAGE to be 39 and 28 kD. The inclusion of proteinase inhibitors during cytosol preparation did not appear to affect the amount of either of the proteins that specifically bound the TCDD photoaffinity ligand. Tissue distribution studies indicated that both proteins were expressed in all tissues examined, but not to the same degree. Gill cytosol had the highest levels of both TCDD-binding proteins. Comparisons of photoaffinity-labeled cytosols from male and female gonads indicated that while the 28 kD protein did not differ significantly, females had significantly more of the 39 kD protein. Oocytes express high levels of the 39 kD protein, but almost none of the 28 kD protein. The kidney contained predominantly the 28 kD protein.

To our knowledge, this is the first report of proteins from a marine invertebrate that specifically bind TCDD. Analogous binding proteins have not been observed in other invertebrates (Hahn *et al.*, 1994). Those that have been reported in vertebrates span a wide range of sizes (95–135 kD in mammals and 105–146 kD in fish), all much larger than the proteins identified in the hard-shell clam. Studies are under way in our laboratory to further characterize these proteins and their cellular roles.

## 3.5 THE IDENTIFICATION OF TRANSFORMING GENES IN BIVALVE TUMOR DNA

The coordinate occurrence of high incidence of ovarian cancer in human populations in the same areas suggests that although bivalves and humans are widely divergent phylogenetically, the gonadal tumors may share a common etiology. Ovarian carcinoma is now one of the leading causes of death due to cancer among women (Barber, 1991) and the molecular basis of these tumors is now an active field of investigation. The currently accepted theory of carcinogenesis is that cell transformation results from a series of gene alterations in primarily two classes of genes, cellular oncogenes and tumor

**Table 3.2** Cellular oncogene alteration in tumors of the reproductive system

| c-Onc | Reference | Tumor | Mechanism |
|---|---|---|---|
| *c-myc* | DiCioccio and Piver, 1992 | Cervical cancer | Amplification, rearrangement |
| c-K-*ras* | DiCioccio and Piver, 1992 | Ovarian tumors | Rare amplification, point mutation |
| c-K-*ras* | DiCioccio and Piver, 1992 | Endometrial | Point mutation adenocarcinoma |
| N-*ras* | DiCioccio and Piver, 1992 | Endometrial | Point mutation adenocarcinoma |
| *HER-2/neu* | Berchuck *et al.*, 1990 | Breast, ovarian cancer | Overexpression, amplification |
| *p53* | Mazars *et al.*, 1991 | Ovarian cancer | Allelic loss, mutation |
| *otu* | Steinhauer *et al.*, 1989 | *Drosophila* ovarian tumor gene | Expression necessary for for germ line development recessive alleles – uncontrolled proliferation, failure to differentiate |

suppressor genes (Table 3.2). It has been well established that both the structure and function of these genes have been highly conserved throughout evolution. Recent studies have connected the overexpression of *HER-2/neu* with advanced stages of epithelial ovarian cancer and poor prognosis (Berchuck *et al.*, 1990). Activation of other proto-oncogenes (K-*ras*, H-*ras*, c-*myc*) have been detected both in ovarian tumors and in established ovarian tumor cell lines (DiCioccio and Piver, 1992). In addition, recent studies of the *p53* tumor suppressor gene in ovarian carcinomas found that 36% of those examined contained mutated alleles in this gene which could lead to the loss of oncosuppressive potential (Mazars *et al.*, 1991). These studies in human populations suggest that it may be of interest to examine the roles of homologous genes in bivalve tumors.

In order to test whether activated oncogenes were present in the clam tumors, DNA was extracted from tumor tissue of both soft-shell and hard-shell clams from eastern Maine and the Indian River, FL, respectively, as described previously (Van Beneden *et al.*, 1993). DNAs from both clam species were co-transfected with the plasmid $pSV_2neo$ into NIH3T3 cells using a calcium phosphate precipitation procedure (Van Beneden *et al.*, 1990, 1993) modified from that of Graham and van der Eb (1973). Cells were assessed for transformation/tumorigenicity in the standard focus assay, colony selection assay and the nude mouse assay (Blair *et al.*, 1982). DNA isolated from cells which were transfected with DNA from an advanced tumor in *Mya arenaria* and by both an early and an advanced tumor from *Mercenaria spp.* were able to transform NIH3T3 cells in both primary (Van Beneden *et al.*, 1993) and secondary (Table

**Table 3.3** Secondary transfection of NIH3T3 cells with DNA from soft-shell (*Mya arenaria*) and hard-shell clams (*Mercenaria spp.*)

| DNA source | Standard focus assay (no. foci/plate) | Nude mouse assay No. tumors/no. mice | Weeks to onset |
|---|---|---|---|
| *Mya arenaria* | | | |
| Reference animal | 0.09 | 2/6 | 4 |
| Early tumor | 1.9 | 0/5 | — |
| Advanced tumor | 1.0 | 5/12 | 6.6 |
| *Mercenaria spp.* | | | |
| Reference animal | 0.6 | 0/6 | — |
| Early tumor | 0.6 | 2/9 | 6 |
| Advanced tumor | 2.4 | 0/7 | — |

*Source*: Data are combined from Van Beneden *et al.* (1993) and Van Beneden (1994).

3.3) cycles of transfection. It should be noted, however, that background activity in control samples was high in the *M. arenaria* reference animal in the nude mouse assay and in the standard focus assay of the DNA from the *Mercenaria* reference animal. Scatter in the focus assay data from the early tumors may be due in part to the presence of mixed normal and tumor tissue in the gonads from the early stage tumors. The high background in the nude mouse assay is a sporadic problem with the assay. These results suggest that a transforming gene(s) may be present in the clam tumor DNA whose structure/function is sufficiently conserved that it remains active in a mammalian cell line. The identity of this gene(s) is still unknown and is actively under investigation.

## 3.6 SEASONAL AND GENDER VARIATION

Hesselman *et al.* (1988) described a seasonal variation in the incidence of germinomas in *Mercenaria* from the Indian River, FL, with the highest incidence occurring during the summer months. This is shortly after the spawning season and at a time when the clams can be stressed by high water temperatures. It is unclear whether the high incidence during this time is associated with either of these two factors. A second observation was that nearly two-thirds of the clams with gonadal tumors were females. In addition, male tumors appeared to be limited to the gonad while some females developed an advanced stage of neoplasia in which tumor cells infiltrated muscle, connective tissue and hemolymph vessels. These results were supported by recent findings (Bert *et al.*, 1993) that the disease occurred at a lower frequency in males and that the percentage of males with the neoplasm decreased with more advanced stages of proliferation. The low frequency of neoplasia in males relative to females together with the high frequency of individual males in the early stage of the disease suggests that either the males are more susceptible and die in the early stages or that they are more resistant to the proliferative stages of the disease.

In our laboratory, we found that the 39 kD TCDD-binding protein was

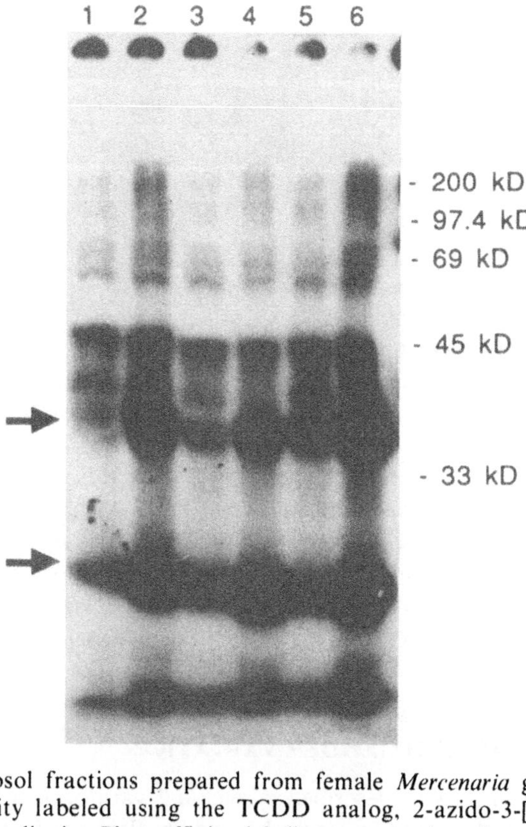


**Figure 3.3** Cytosol fractions prepared from female *Mercenaria* gonadal tissue were photoaffinity labeled using the TCDD analog, 2-azido-3-[$^{125}$I]-iodo-7,8,-dibromodibenzo-p-dioxin. Photoaffinity labeling was conducted according to the procedure described by Poland *et al.* (1986) with minor modifications. The odd-numbered samples were labeled in the presence of cold competitor (300-fold molar excess of 2,3,7,8-tetrachlorodibenzofuran). During labeling, the concentration of the photoaffinity ligand was 0.6 nM. The proteinase inhibitors leupeptin (5 μg/ml) and pepstatin (5 μg/ml) were included during the preparation of cytosols used for samples 1–4. The cytosol proteins were separated by polyacylamide gel electrophoresis on a 12% tris-glycine continuous gel. The arrows to the left of the photograph indicate the positions of the two proteins (*upper arrow*, 39 kD; *lower arrow*, 28 kD) identified in *M. mercenaria* which specifically bound the TCDD analog.

present at higher levels in female clam gonads than in male clams. Whether the higher level of expression in the female gonad is related to the higher incidence of gonadal tumors in female clams is still unclear.

The results of our study raise several questions about the role of these proteins in hard-shell clams and TCDD-binding proteins in general. The presence of two proteins and their differential tissue expression suggests that they could have unique cellular roles. In vertebrates, the ligand-Ah receptor-ARNT complex acts to alter transcription of various genes (Sutter and Greenlee, 1992; Hahn *et al.*, 1993). Further studies are planned to investigate the possible role of the clam dioxin-binding proteins as transcriptional enhancers of gene expression.

## 3.7 GENETIC ELEMENTS

While the hypothesis that herbicide exposure may have a role in the etiology of these gonadal tumors in both soft- and hard-shell clams is intriguing it is also controversial and, at this time, still far from proven. While there is convincing correlative evidence for the role of herbicides in tumor formation in the Maine soft-shell clams, it is not as clear in the Florida hard-shell clam populations. Other important factors are beginning to emerge. The genetic contributions to this disease in *Mercenaria* from this region have been addressed by Bert *et al.* (1993). In the region where a notably high incidence of gonadal neoplasm was reported by Hesselman *et al.* (1988), two species (*M. mercenaria* and *M. campechiensis*) occur sympatrically (Dillon and Manzi, 1989). Hybridization occurs readily between them and hybrids may be distinguished by allozyme patterns (Dillon and Manzi, 1989) mitochondrial genotype (Brown, 1989) and shell morphology (Abbott, 1974; Dillon and Manzi, 1989; Menzel, 1989). Bert *et al.* (1993) report that gonadal tumors occurred two to three times as often in the hybrid animals relative to either parental species. This parallels the case of melanoma formation in the *Xiphophorus* hybrids which are susceptible to both spontaneous melanoma formation and sensitive to tumor induction by chemical carcinogens (Anders, 1989). Bert *et al.* (1993) proposed that hybridization in *Mercenaria* may have disrupted genetic mechanisms that are involved in cell growth and differentiation but not the regulation of cellular proliferation. Regardless of the etiological factors, the genetic elements involved must be complex. These observations do not rule out the possibility that the hybrids may be more susceptible to environmental contaminants.

## 3.8 CONCLUSION

Preliminary investigations of the molecular mechanisms of gonadal tumor formation in two bivalve species have proceeded along several lines and now pose as many questions as they have answered. This study represents our initial efforts at understanding the link between human and environmental health. The coordinate occurrence of high incidence of ovarian cancer in human populations in the same areas suggests that although bivalves and humans are widely divergent phylogenetically, these tumors may share a common etiology. The results of our preliminary transfection studies indicate that clam tumor DNA is able to transform NIH3T3 cells. Although the identity of this gene(s) has not yet been determined, our data suggest that it may be functionally conserved. In a second line of investigation, two proteins have been detected which specifically bind a dioxin analog. We have proposed that these proteins may be candidates for an invertebrate analog of the clam Ah receptor. In spite of these intensive investigations, the etiology of these tumors is still unknown.

*ACKNOWLEDGMENTS*

I wish to thank G. Gardner and Dr N. Blake for providing samples and histopathology of tumor bearing clams, Dr G. Clark and D. Brown for their work on the photoaffinity binding studies and Dr Z. Fan and L. Lynch for their technical assistance in the transfection studies. Supported in part by a grant from the NIH (R01 RR08774-01).

## REFERENCES

Abbott, R. T. (1974) *American Sea Shells.* Van Nostrand Reinhold, New York.

Anders, F. (1989) The Mildred Scheel 1988 Memorial Lecture: A biologist's view of human cancer, in: *Modern Trends in Human Leukemia*, VIII, Springer Verlag Press, Heidelberg, pp. XXIII–XLV.

Barber, H. R. K. (1991) New frontiers in ovarian cancer diagnosis and management. *Yale J. Biol. Med.*, **64,** 127–41

Baumann, P. C., Smith, W. D. and Ribick, M. (1982) Hepatic tumor levels and polynuclear aromatic hydrocarbon levels in two populations of brown bullheads (*Ictalurus nebulosus*), in *Polynuclear Aromatic Hydrocarbons: Physical and Biological Chemistry*, (eds M. Cooke, A. J. Dennis and G. L. Fisher), Battelle Press, Columbus, OH, pp. 93–102.

Berchuck, A., Kamel, A., Whitaker, R. *et al.* (1990) Overexpression of *HER-2/neu* is associated with poor survival in advanced epithelial ovarian cancer. *Cancer Res.*, **50,** 4087–91.

Bert, T. M., Hesselman, D. M., Arnold, W. S. *et al.* (1993) High frequency of gonadal neoplasia in a hard clam (*Mercenaria spp.*) hybrid zone. *Mar. Biol.*, **117,** 97–104.

Blair, D. G., Cooper, C. S., Oskarsson, M. K. *et al.* (1982) New method for detecting cellular transforming genes. *Science*, **218,** 1122–25.

Black, J. J. (1983) Field and laboratory studies of environmental carcinogenesis in Niagara River fish. *J. Great Lakes Res.*, **9,** 326–34.

Brown, B. L. (1989) Population variation in the mitochondrial DNA of two marine organisms: the hard-shell clam *Mercenaria spp.* and the killifish, *Fundulus heteroclitus.* PhD dissertation, Old Dominion University, Norfolk, VA.

Brown, D. J., Van Beneden, R. J. and Clark, G. C. (1995) Identification of two binding proteins for planar halogenated aromatic hydrocarbons in the hardshell clam, *Mercenaria mercenaria. Arch. Biochem. Biophys.*, **319,** 217–24.

Burbach, K. M., Poland, A. and Bradfield, C. A. (1992) Cloning of the Ah-receptor cDNA reveals a distinctive ligand-activated transcription factor. *Proc. Natl. Acad. Sci.*, **89,**8185–89.

DiCioccio, R. A. and Piver, M. S. (1992) The genetics of ovarian cancer. *Cancer Invest.*, **10,** 135–41.

Dillon, R. T. Jr and Manzi, J. J. (1989) Genetics and shell morphology in a hybrid zone between the hard clams *Mercenaria mercenaria* and *M. campechiensis. Mar. Biol.*, **100,** 217–22.

Gardner, G. R. (1993) Chemically induced histopathology in aquatic invertebrates, in *Pathobiology of Marine and Estuarine Organisms*, (eds J. A. Couch and J. W. Fournie), CRC Press, Boca Raton, FL, pp. 359–91.

Gardner, G. R., Yevich, P. P., Harshbarger, J. C. and Malcolm, A. R. (1991a) Carcinogenicity of Black Rock Harbor sediment to the eastern oyster and trophic transfer of Black Rock Harbor carcinogens from the blue mussel to the winter flounder. *Environ. Health Perspec.*, **90,** 53–66.

Gardner, G. R., Yevich, P. P., Hurst, J. *et al.* (1991b) Germinomas and teratoid siphon anomalies in softshell clams, *Mya arenaria*, environmentally exposed to herbicides. *Environ. Health Perspect*, **90,** 43–51.

Graham, F. L. and van der Eb, A. J. (1973) A new technique for the assay of infectivity of human adenovirus 5 DNA. *Virology*, **52,** 456–67.

Hahn, M. E., Lamb, T. M., Schultz, M. E. *et al.* (1993) Cytochrome p450A induction and inhibition by 3,3′,4,4′-tetrachlorobiphenyl in an Ah receptor-containing fish hepatoma cell line (PLHC-1). *Aquatic Tox.*, **26,** 185–208.

Hahn, M. E., Poland, A., Glover, E. and Stegeman, J. J. (1994) Photoaffinity labeling of the Ah receptor: phylogenetic survey of diverse vertebrate and invertebrate species. *Arch. Biochem. Biophys.*, **310,** 218–28.

Hankinson, O. (1993) Research on the aryl hydrocarbon (dioxin) receptor is primed to take off. *Arch. Biochem. Biophys.*, **300,** 1–5.

Hanson, D. J. (1991) Dioxin toxicity: new studies prompt debate, regulatory action. *Chem. Engineering News*, pp. 7–14.

Harshbarger, J. C. and Clark, J. (1990) Epizootiology of neoplasms in bony fish of North America. *Sci. Total Environ.*, **94,** 1–32.

Hayes, H. M., Tarone, R. E., Casey, H. W. and Huxsoll, D. L. (1990) Excess of seminomas observed in Vietnam Service U.S. military working dogs. *J. Natl Cancer Inst.*, **82,** 1042–46.

Hayes, H. M., Tarone, R. E., Cantor, K. P. *et al.* (1991) Case-control study of canine malignant lymphoma: positive association with dog owner's use of 2,4-dichlorophenoxyacetic acid herbicides. *J. Natl. Cancer Inst.*, **83,** 1226–31.

Hesselman, D. M., Blake, N. J. and Peters, E. C. (1988) Gonadal neoplasms in hardshelled clams, *Mercenaria spp.*, from the Indian River, Florida: occurrence, prevalence and histopathology. *J. Invert. Pathol.*, **52,** 436–46.

Krahn, M. M., Rhodes, L. D., Myers, M. S. *et al.* (1986) Associations between metabolites of aromatic compounds in bile and occurrence of hepatic lesions in the English sole (*Parophyrus vetulus*) from Puget Sound, Washington. *Arch. Environ. Contam. Toxicol.*, **15,** 61–7.

Laemmli, U. K. (1970) Cleavage of structural proteins during the assembly of the head of bacteriophage T4. *Nature*, **227,** 680–5.

Landers, J. P. and N. J. Bunce. (1991) The Ah receptor and the mechanism of dioxin toxicity. *Biochem. J.*, **276,** 273–87.

Lorenzen, A. and Okey, A. B. (1990) Detection and characterization of [$^3$H] 2,3,7,8-tetrachlorodibenzo-p-dioxin binding to Ah receptor in a rainbow trout hepatoma cell line. *Toxicol. Appl. Pharmacol.*, **106,** 53–62.

Matthews, J. (1992) Female veterans seek answers: Vietnam cancer risks in question. *J. Natl. Cancer Inst.*, **84,** 1462–3.

Mazars, R., Pujol, P., Maudelonde, T. *et al.* (1991) *p53* mutations in ovarian cancer: a late event? *Oncogene*, **6,** 1685–90.

Menzel, R. W.(1989) The biology, fishery and culture of quahog clams, *Mercenaria,* in *Clam Mariculture in North America,* (eds J. J. Manzi and M. Casagna), Elsevier Press, Amsterdam, pp. 201–42.

Murchelano, R. A. and Wolke, R. E. (1985) Epizootic carcinoma in the winter flounder, *Pseudopleuronectes americanus. Science*, **288,** 587–9.

Pittenger, C. A., Buikema, A. L. Jr, Horner, S. G. and Young, R. W. (1985) Variation on tissue burdens of polycyclic aromatic hydrocarbons in indigenous and relocated oysters. *Environ. Toxicol. Chem.*, **4,** 379–87.

Pittenger, C. A., Buikema, A. L. Jr and Falkingham, J. O. III. (1987) *In situ* variations in oyster mutagenicity and tissue concentrations of polycyclic aromatic hydrocarbons. *Environ. Toxicol. Chem.*, **6,** 51–60.

Poland, A. and Knutson, J. C. (1982) 2,3,7,8-tetrachlorodibenzo-*p*-dioxin and related halogenated aromatic hydrocarbons: examination of the mechanism of toxicity. *Ann. Rev. Pharmacol. Toxicol.*, **22,** 517–54.

Poland, A., Glover, E., Ebetino, F. H. and Kende, A. S. (1986) Photoaffinity labeling of the Ah receptor. *J. Biol. Chem.*, **261,** 6352–65.

Poland, A., Glover, E. and Kende, A. S. (1976) Stereospecific high affinity binding of 2,3,7,8-tetrachlorodibenzo-p-dioxin by hepatic cytosol. *J. Biol. Chem.*, **251,** 4936–46.

Rier, S. E., Martin, D. C., Bowman, R. E. *et al.* (1993) Endometriosis in rhesus monkeys (Macaea mulatta) following chronic exposure to 2,3,7,8-tetrachlorodibenzo-p-dioxin. *Fund. Appl. Toxicol.*, **21,** 433–41.

Riggan, W. B., Creason, J. P., Nelson, W. C. *et al.* (1987) *US Cancer Mortality Rates and Trends, 1950–1979*, vol. IV: *Maps.* EPA/600/1–83/015e, US EPA, Health Effects Research Laboratory, Research Triangle Park, NC.

Roberts, L. (1991) Dioxin risks revisited. *Science*, **251,** 624–6.

Schmidt, K. F. (1992) Dioxin's other face: portrait of an 'environmental hormone'. *Science News*, **141,** 24–7.

Silbergeld, E. K. and Gasiweicz, T. A. (1989) Dioxins and the Ah receptor. *Am. J. Indust. Med.*, **16,** 455–74.

Smith, C. E., Peck, T. H., Klauda, R. H. and McLaren, J. B. (1979) Hepatomas in Atlantic Tomcod (*Microgadus tomcod*) collected in the Hudson River Estuary, NY. *J. Fish Dis.*, **2,** 313–19.

Steinhauer, W. R., Walsh, R. C. and Kalfayan, L. J. (1989) Sequence and structure of the *Drosophila melanogaster* ovarian tumor protein. *Mol. Cell Biol.*, **9,** 5726–32.

Sutter, T. R. and Greenlee, W. F. (1992) Classification of members of the Ah gene battery. *Chemosphere*, **25,** 223–6.

Van Beneden, R. J. (1994) Molecular analysis of bivalve tumors: models for environmental and genetic interactions. *Environ. Health Pers.*, **102,** (Suppl. 12), 81–83.

Van Beneden, R. J., Henderson, K. W., Blair, D. G. *et al.* (1990) Oncogenes in hematopoietic and hepatic fish neoplasms. *Cancer Res.* (Suppl.), **50,** 5671s–4s.

Van Beneden, R. J., Gardner, G. R., Blake, N. J. and Blair, D. G. (1993) Implications for the presence of transforming genes in gonadal tumors in two bivalve mollusk species. *Cancer Res.*, **53,** 2976–9.

Whitlock, J. P. (1990) Genetic and molecular aspects of 2,3,8-tetrachlorodibenzo-p-dioxin action. *Ann. Rev. Pharmacol. Toxicol.*, **30,** 251–77.

Zeise, L., Huff, J. E., Salmon, A. G. and Hooper, N. K. (1990) Human risks from environmental exposure to 2,3,7,8-tetrachlorodibenzo-p-dioxin and hexachloro-dibenzo-p-dioxins. *Adv. Mod. Environ. Toxicol.*, **17,** 293–342.

Yevich, P. P. and Barry, M. M. (1967) Ovarian tumor in the quahog (*Mercenaria mercenaria*). *J. Invert. Pathol.*, **14,** 266–7.

Yevich, P. P. and Barszcz, C. A. (1977) Neoplasia in soft shell clams (*Mya arenaria*) collected from oil-impacted sites. *Ann. NY Acad. Sci.*, **298,** 409–26.

Yuspa, S. H. and Poirier, M. C. (1988) Chemical carcinogenesis: from animal models to molecular models in one decade. *Adv. Can. Res.*, **50,** 25–70.

# 4 *Emerging issues: the effects of endocrine disrupters on reproductive development**

L. EARL GRAY JR, EMILY MONOSSON and
WILLIAM R. KELCE

## 4.1 INTRODUCTION

The effects of 'endocrine disrupters' on the health of human and wildlife populations are currently receiving a great deal of interest from the popular press and the scientific community. At the present, scientists are grappling with research needs in this new area and the need for new test protocols to screen for endocrine effects. The impetus for this new approach began at a Work Session in July of 1991 on 'Chemically Induced Alterations in Sexual Development: the Wildlife/Human Connection' attended by a multidisciplinary group of experts (Colborn and Clement, 1992). A consensus statement from this Work Session (Colborn and Clement, 1992: 1–6) concluded that 'Many compounds introduced into the environment by human activity are capable of disrupting the endocrine system of animals, including fish, wildlife, and humans. The consequences of such disruption can be profound because of the crucial role hormones play in controlling development.' In addition, one portion of the consensus statement indicated that many wildlife populations were already affected by such compounds, with effects (relevant to this chapter) including demasculinization and feminization of male fish, birds and

*The research described in this article has been reviewed by the Health Effects Research Laboratory, US Environmental Protection Agency, and approved for publication. Approval does not signify that the contents necessarily reflect the view and policies of the Agency nor does mention of trade names or commercial products constitute endorsement or recommendation for use.

*Interconnections Between Human and Ecosystem Health.* Edited by Richard T. Di Giulio and Emily Monosson. Published in 1996 by Chapman & Hall, London. ISBN 0 412 62400 1.

mammals and defeminization and masculinization of female fish and birds.

Among these chemicals are persistent, bioaccumulative, organohalogen compounds that include pesticides and industrial chemicals, other synthetic products and some metals (Table 4.1). Laboratory studies corroborate the abnormalities of reproductive development observed in the field and, in some cases, provide mechanisms to explain the effects in wildlife. The Work Session also 'estimated with confidence' that developmental impairments in humans have resulted from exposure to synthetic endocrine disrupters that are present in the environment. Recent findings have contributed to these concerns; for example, it has been suggested that *in utero* exposure to environmental estrogens or chemicals like TCDD may be responsible for the reported 50 % decline in sperm counts over the past 50 years and the apparent increase in cryptorchid testes, testicular cancer and hypospadias (Carlsen *et al.*, 1992; Giwercman *et al.*, 1993). In females, exposure to endocrine disrupting chemicals during development could contribute to earlier age at puberty (Hannon *et al.*, 1987), increased incidences of endometriosis (Koninckx *et al.*, 1994) and breast cancer (Davis *et al.*, 1993). Concerns about the effects of exposure to endocrine disrupters during development led the National Academy of Sciences to release a report on 'Pesticides in the Diets of Infants and Children', which suggested that the young are a special concern with respect to pesticide exposures (National Research Council, 1993).

To date, the discussion of 'endocrine disrupters' has been focused on toxicants reported to possess estrogenic activity, with little consideration given to other mechanisms of endocrine toxicity; mechanisms that, in fact, may be of equal or greater concern. In addition, there has been a great deal of misinformation communicated in the press and in scientific journals on issues concerning endocrine disrupters; for example, non-estrogenic chemicals (p, p′-DDE) are repeatedly reported to be estrogenic. There is a lack of appreciation for the fact that many endocrine disrupters (i.e. TCDD, p, p′DDE, vinclozolin metabolite M2) are very potent reproductive toxicants. In addition, there has been a tendency to dismiss the wildlife data as correlative, ignoring examples of clear cause and effect relationships between chemical exposure and reproductive alterations (e.g. DDT metabolites effects in birds, PCB effects in fish and environmental estrogen effects in domestic animals). There is a lack of recognition of the fact that subtle, low dose reproductive effects seen in laboratory studies with endocrine disrupters will be difficult, if not impossible, to detect in typical epidemiological studies because of high variability normally seen in human reproductive function (e.g. time to fertility, fecundity and sperm measures) and the delayed appearance of the reproductive lesions. There is a lack of appreciation for the complexity of the endocrine system, which leads to a failure to comprehend the multiple mechanisms by which a single chemical can alter the endocrine milieu, to say nothing of the complex endocrine alterations induced by a mixture like DDT or the PCBs and their metabolites.

**Table 4.1** Endocrine disrupters*

**A. Reproductive endocrine disrupters**

1. Environmental estrogens – estrogen receptor mediated
   - Methoxychlor
   - Chlordecone
   - Some PCBs
   - β-isomer of lindane
   - o,p′-DDT
   - Bisphenolic compounds – octyl/nonylphenols/APEs
   - 6-nitro-1,3,8 trichlorodibenzofuran
2. Environmental antiestrogens
   - Dioxin (downregulates estrogen receptor at high doses)
   - pp-DDT, DDE (increase degradation of estrogen in birds)
   - Endosulfan (inhibits vitellogenesis in fish)
3. Environmental antiandrogens/androgens
   - Androgen receptor mediated
   - Vinclozolin (binds to the androgen receptor)
   - Procymidone
   - DDE
   - Synthetic environmental estrogens (nonylphenol, methoxychlor metabolite)
   - American dwarf palm extract
   - Kraft mill effluent
4. Toxicants that reduce serum levels of a steroid hormone
   - Fenarimol and other fungicides (two classes)
   - Dioxin (inhibits T synthesis in adult male rat)
   - Aroclor 1254 (decreases serum progesterone in pregnant rats)
   - R11 (insect repellent *in vitro*)
   - Endosulfan (male rat – reduced T, FSH and LH)
   - Mirex (blocked ovulation in rat)
   - Dibutyl phthalate (lowered serum P4 in pregnant females)
   - Lead (monkey – altered menstrual cycles)
5. Toxicants that affect reproduction primarily via the CNS
   - Thiram, metam sodium and other dithiocarbamate pesticides (inhibit the LH surge and ovulation)
   - Amitraz and chlordimeform (inhibits the LH surge)
   - Carbon disulfide (behavioral)
   - Chlorpyriphos and some other OPs
   - Methanol
   - Manganese (human impotence – may be a CNS effect)
6. Other toxicants that alter hormonal status
   - Dibutyl phthalate (increase FSH)
   - Cadmium injections (″)
   - Benomyl/carbendazim (″)
   - Benzidine-based dyes (″)
   - Vinylcyclohexene (kills oocytes)
   - DMBA, benzo(A)pyrene (PAHs) (kills oocytes)
   - Vinclozolin, iprodione, linuron (cause Leydig cell/ovarian tumors, proposed endocrine mechanisms)
   - DBCP (kills germ cells and alters sex differentiation)

**Table 4.1** Continued

| | |
|---|---|
| **B.** | **Antithyroid endocrine disrupters** |
| | 1. Many thiocarbamide and sulfonamide-based pesticides ETU, EDBC linuron and etc. |
| | 2. PCBs |
| | 3. Dioxin |
| | 4. PCDFs |
| | 5. The herbicide nitrofen |
| | 6. Radiation – thyroid tumors |
| | 7. Lead |
| | 8. Mirex |
| | 9. Phthalic acid esters |
| | 10. PBBs |
| | 11. Hexachlorobenzene |
| **C.** | **Adrenal endocrine disrupters** † |
| | 1. Vinclozolin and related fungicides |
| | 2. Mirex |
| | 3. Aniline dyes |
| | 4. Carbon tetrachloride |
| | 5. Chloroform |
| | 6. o,p′-DDD, DDE |
| | 7. Dimethylbenzathracene |
| | 8. Methanol, ethanol |
| | 9. Fungicides of the ketoconazole class |
| | 10. Nitrogen oxides |
| | 11. Parathion |
| | 12. PBBs, PCBs |
| | 13. Toxaphene |
| | 14. Urethane |

* This list is not intended to be all-encompassing, but rather it is intended to show that the chemicals that have been shown to alter the endocrine system are not limited to 'environmental estrogens'. Pesticides and toxic substances can attack *multiple* sites in the reproductive system. In addition, many toxicants have also been shown to alter thyroid and adrenal function as well. The fact that a toxicant is on this list indicates that it interacts with the endocrine system in some manner, but does *not* indicate that humans are exposed to toxic dosage levels.
† See review by Colby (1988).

In the remainder of this chapter we will present new information on the effects of endocrine disrupters on reproductive development. Data will be presented that describe how exposure to environmental estrogens (natural and synthetic), environmental antiandrogens, TCDD and PCBs during critical stages of life results in abnormal sex differentiation and pubertal development. The discussion will include selected examples of chemically induced alterations of reproductive function from human, laboratory and wildlife studies. In addition, the potency of some of these environmental endocrine disrupters will be compared to the steroid hormones and drugs. One final issue that needs to receive more attention is the subject of screening toxicants for hormonal activity. Opinions regarding the best methods for screening are varied, ranging from exclusively using *in vitro* systems to exclusively using *in vivo* screens. For the purposes of discussion, we provide an example of a comprehensive investigational strategy intended to identify not only the chemical/metabolite of concern, but also provide information as to the mechanism responsible for the phenotypic effects at the end of this chapter.

## 4.2 SEXUAL DIFFERENTIATION: BACKGROUND

In humans and rodents, exposure to hormonally active chemicals during sex differentiation can produce morphological pseudohermaphrodism (Gray, 1992; Schardein, 1993). For example, the potent estrogen DES causes urogenital malformations and cancer in the reproductive tracts of humans and rodents. In addition, exposure to the androgenic substances Danazol or methyltestosterone masculinizes human females (i.e. 'female pseudohermaphroditism'). Progestins act both as androgen antagonists, demasculinizing males such that they display ambiguous genitalia with hypospadias (Schardein, 1993), and as androgen agonists, masculinizing females. Laboratory studies demonstrate that these chemicals also produce alterations of sex differentiation in rodents that closely resemble those seen in humans (Gray, 1992). The drug aminoglutethimide, which alters steroid hormone synthesis in a manner identical to many fungicides, also masculinizes human females following *in utero* exposure.

Exposure to toxic substances during critical developmental periods can alter functional reproductive development in humans. *In utero* exposure to PCBs and PCDFs has been associated with reproductive alterations in boys, increased stillbirths, low birth weight, malformations and IQ and behavioral deficits (Guo *et al.*, 1993). All of the above human developmental reproductive toxicants alter reproductive development in laboratory animals in a predictable manner, because the basic mechanisms underlying sexual differentiation are homologous in all mammals, although the timing of certain events varies. Hence, chemicals that have adverse effects on reproductive development in rodents and other mammals, including wildlife and domestic animal species, should be considered as potential human reproductive toxicants as well. Reproductive alterations in avian, reptilian and amphibian species also should be cause for concern because considerable homology exists between the reproductive physiology of these different classes of vertebrates.

## 4.3 NORMAL SEXUAL DIFFERENTIATION

A wealth of information exists regarding the role of hormones in sex differentiation in humans and other mammals because even the most severe alterations of this process are not lethal. Normally the development of the male or female phenotype from an indifferent state entails a complex series of events (Wilson,

1978). Genetic sex is determined at fertilization and this normally governs the expression of the 'male factor' and the subsequent differentiation of gonadal sex. Prior to sex differentiation, the embryo has the potential to develop a male or female phenotype. Following gonadal sex differentiation, testicular secretions induce differentiation of the male duct system and external genitalia. The development of phenotypic sex includes persistence of either the Wolffian (male) or Mullerian (female) duct system, and differentiation of the external genitalia and the central nervous system (CNS). Other organ systems, like the liver, are sexually 'imprinted' as well. The male phenotype arises due to the action of testicular secretions, testosterone and Mullerian inhibiting substance. In the human embryo, the onset of testosterone synthesis by the testis occurs 65 days after fertilization. Testosterone induces the differentiation of the Wolffian duct system into the epididymis, vas deferens and seminal vesicles, while its metabolite, 5-alpha-dihydrotestosterone (DHT), induces the development of the prostate and male external genitalia. It has generally been held that, in the absence of these secretions, the female phenotype is expressed (whether or not an ovary is present). However, the ovary secretes estrogens during development and recent studies have suggested that this hormone plays an important role in feminization of the female reproductive tract, although this hypothesis is in conflict with the observation of normally differentiated reproductive tracts in homozygous recessive female ERKO (estrogen-receptor knock out) mice that are apparently completely devoid of estrogen receptors (Korach, 1994). The external genitalia of human male and female fetuses (18–22 weeks' gestation) both express androgen receptors, but male external genitalia lacked estrogen receptors, ruling out a direct influence of maternal estrogen on male genital development (Kalloo *et al.*, 1993).

In the CNS, testosterone is aromatized (via the steroidogenic enzyme aromatase) to estradiol and reduced via 5-alpha reductase to DHT. It has been suggested for some species that all three hormones (T, DHT and $E_2$) play a role in the masculinization of the CNS. In the rat, mouse and hamster aromatization of testosterone to estradiol mediates CNS sex differentiation. For example, in the hypothalamus the sexually dimorphic nucleus in the preoptic area (SDN-POA) of the rat is several-fold larger in male rats than in female rats due to the perinatal influence of estradiol. Although similar morphological sex dimorphisms have been reported for the human CNS, the role of estrogens in CNS sex differentiation process is unclear, but the androgenic (T and/or DHT) pathway is essential for CNS sex differentiation in the rhesus monkey (McEwen, 1980).

Initially, sex differences in lateralization of the human cortex and in the preoptic area of the hypothalamus were reported by Swaab and Fliers (1985). They found that the preoptic area of the hypothalamus is 2.5 times larger in men than in women and contains 2.2 times as many cells, an effect that is reminiscent of the sex dimorphism present in the hypothalamus of the rat (Gorski *et al.*, 1978). In addition, a quantitative analysis of the volume of four cell groups in the preoptic-anterior hypothalamic area (PO-AHA) and the supraoptic nucleus of 22 age- matched men and women found gender-related differences in two of four cell groups in the POAHA. The interstitial nuclei of the anterior hypothalamus (INAH), INAH-2 and INAH-3, were at least twice as large in the male brain as in the female brain (Allen *et al.*, 1989). In the rat, this area of the hypothalamus contains a neuronal network that is essential for gonadotropin release and sexual behavior (Leranth *et al.*, 1985). Le Vay (1991) not only found a similar sexual dimorphism in INAH-3, but he also reported that homosexual men have female-like INAH-3 structures, implying a functional component to this CNS sex dimorphism.

In man, there are a number of genetic errors involving sex-determining mechanisms including: complete and incomplete sex reversals (XX males and XY females); sex chromosome anomalies (Polani, 1981); single gene defects coding for a defective steroidogenic enzyme, which lead to reduced synthesis of sex steroids (20,22-desmolase; 17-ketosteroid reductase; and 5 alpha-reductase deficiency); defective steroid receptor, resulting in abnormal handling of androgens in the target tissues (complete testicular feminization; Reifensten syndrome); and various other genetic defects (LH deficiency and lack of responsiveness to human chorionic gonadotropin). In all, more than two dozen different genes regulate sexual development in man, and each mutation has a profound and often unique effect on the sexual phenotype (Polani, 1981). When similar congenital reproductive abnormalities have been detected in laboratory species, they have dramatically facilitated our understanding of genetic errors of steroid metabolism and receptor function in man. For this reason, rodent and other models have great utility for evaluating the potential of xenobiotics to alter human reproductive development. In summary, effects of exposure to endocrine disrupters during sex differentiation are of special concern for a number of reasons; this process is very sensitive to the effects of relatively low doses of endocrine disrupters; the effects are irreversible; the system is 'imprinted' by the initial hormonal environment; functional alterations of the sex differentiation are often not apparent until after puberty, or even later in life; and the abnormalities, which include malformations and infertility, cannot be predicted from the transient alterations in hormone levels produced by similar exposure in adult animals. For these reasons, developmental reproductive toxicity data are often critical in the assessment of non-cancer health effects of endocrine disrupters.

## 4.4 MECHANISMS OF ENDOCRINE TOXICITY: ESTROGENS

### *4.4.1 ESTROGENIC DRUGS: MODELS FOR ENVIRONMENTAL ESTROGENS*

DES provides a grim example of how *in utero* exposure to a potent endocrine disrupter with estrogenic activity can alter reproductive development in humans. Although a few cases of masculinized females and demasculinized males were noted in the late 1950s, most of the effects of DES were not apparent until after the children attained puberty. Transplacental exposure of the developing fetus to DES at critical periods leads to similar abnormalities of the urogenital tracts of both rodents and primates. The pathological effects that develop in women as a consequence of *in utero* exposure to the potent estrogenic drug DES are well established. DES causes clear cell adenocarcinoma of the vagina, as well as gross structural abnormalities of the cervix, uterus and fallopian tube. These women are more likely to have an adverse pregnancy outcome, including spontaneous abortions, ectopic pregnancies and premature delivery (Steinberger and Lloyd, 1985). The fact that human male fetuses lack estrogen receptors (Kalloo *et al.*, 1993), coupled with our observations of androgen receptor binding by estradiol and DES *in vitro* at high but toxicologically relevant concentrations, suggests that 'antiandrogenic' effects of DES may be mediated through androgen receptor binding and antagonism of the actions of testosterone and DHT. Some of the pathological effects that develop in fetal males following DES exposure appear to result from an inhibition of androgens (hypospadias, underdevelopment or absence of the vas deferens, epididymis and seminal vesicles) and anti-Mullerian duct factor (persistence of the Mullerian ducts) (McLachlan, 1981; Steinberger and Lloyd, 1985; Schardein, 1993). DES also causes epididymal cysts, hypotrophic testes and infertility in males. Some males have reduced ejaculate volume with reduced numbers of motile sperm, and some also experience difficulty in urination.

In addition, it has been reported that DES can alter sex differentiation of the human brain, in addition to the reproductive tract. However, the effect of DES on human behavioral sex dimorphisms is controversial. Several behavioral alterations have been observed in some DES-daughters (Meyer-Bahlburg *et al.*, 1985; reviewed by Hines and Green, 1991). Meyer-Bahlburg *et al.* (1985) reported that women exposed to DES *in utero* were found to have less well established sex-partner relationships, and to be lower in sexual desire and enjoyment, sexual excitability and coital functioning. In addition, Hines and Shipley (1984) found that DES-exposed women showed a more masculine pattern of cerebral lateralization on a verbal task than did their sisters. Such sex differences in specialization of the two hemispheres of the brain for different types of cognitive processing are well documented in humans, with men tending towards greater left-hemisphere specialization for verbal stimuli than women (McGlone, 1980). However, a recent review of gender-related behavior in women exposed prenatally to DES concluded that these studies did not provide clear-cut differences between unexposed and DES-exposed women (Newbold, 1993). Clearly, more studies are needed to clarify the role of estrogens in human CNS sex differentiation. It is clear that androgens play a more prominent role in this process in non-human primates than in rodents, where estrogenic actions regulate many phases of CNS sex differentiation. In the female rhesus monkey, androgen (T and DHT) administration during prenatal life defeminizes (Pomerantz *et al.*, 1985) and masculinizes (Goy, 1978) some reproductive behaviors and rough-and-tumble play. The evidence also suggests that physical energy expenditure by children during play is similarly influenced by prenatal androgens (Ehrhardt and Meyer-Bahlburg, 1981).

Effects seen in female and male animals exposed *in utero* or neonatally to estrogens resemble some of those seen in DES-exposed sons and daughters. In rodents, after DES exposure the fetal anlagen of both sexes exhibit defective development such that both male and female progeny retain reproductive tract remnants of the opposite sex. In female progeny, the administration of DES to rodents during the perinatal life reduces fertility, produces structural abnormalities of the oviduct, uterus, cervix and vagina (McLachlan, 1981; McLachlan *et al.*, 1982; Newbold, 1993), squamous metaplasia is present in the uterus and part of the cervix (Ennis and Davies, 1982), and the incidence of mammary carcinomas is doubled (Huseby and Thurlow, 1982) in the female offspring. In male mice, perinatal DES causes epididymal cysts, hypospadias, phallic hypoplasia, inhibition of growth and descent of the testes, and underdevelopment or absence of the vas deferens, epididymis and seminal vesicles (McLachlan, 1981). In addition, female structures derived from the Mullerian ducts persist in male mice after *in utero* administration of DES (Newbold and McLachlan, 1985; Newbold *et al.*, 1987). DES-like malformations are obtained in male and female rodents after perinatal administration of several other potent estrogens like estradiol and RU 2858 (Vannier and Raynaud, 1980; Bern *et al.*, 1983).

DES clearly alters sex differentiation of the rodent and avian CNS as well as the reproductive tract. DES masculinizes the CNS of female rodents (male being heterogametic- XY) such that the SDN-POA is larger and male-like, while in avian species DES and other estrogenic substances demasculinize the male (female being heterogametic- ZW).

The administration of antiestrogens is known to alter reproductive development in neonatal rodents. Neonatal exposure to some of these chemicals, like tamoxifen, nafoxidine and clomiphene, causes gross abnormalities of reproductive development (Clark and McCormack, 1977; Chamness *et al.*, 1979). Vaginal opening was accelerated, estrous cycles were absent at 4 months of age, the ovaries and uteri were atrophic and the oviducts showed squamous metaplasia. Female offspring exposed *in utero* to the antiestrogen LY117018 exhibited cleft phallus and oviduct malformations similar to rats treated with DES and estradiol treated rats (Henry and Miller, 1986). Perinatal treatment of female rats with the antiestrogen tamoxifen resulted in permanent anovulatory sterility, although it did not produce estrogen-like structural changes in the SDN-POA (Dohler *et al.*, 1986). In the neonatal female rodent, a number of these 'antiestrogens' actually act as estrogen receptor agonists, rather than antagonists, which confounds the interpretation of these data.

The preceding discussion has presented scientific evidence demonstrating that prenatal exposure to estrogenic chemicals can result in dramatic abnormalities of human reproductive development. The reproductive alterations include morphological and pathological abnormalities that were not apparent until well after childhood. One of these chemicals, DES, may also alter human CNS development as indicated by changes in sexually dimorphic behaviors. Many of the above human pathologies are similar to experimentally induced alterations of sex differentiation in studies using rodents.

### *4.4.2 ENVIRONMENTAL EXTROGENS – PHYTOESTROGENS AND MYCOTOXINS*

#### (a) Plant estrogens

Although DES and other potent estrogenic drugs produce dramatic alterations of sex differentiation, these chemicals are rarely found in the environment. However, plant estrogens and fungal toxins, pesticides and toxic substances, including a few PCBs, have also been shown to alter sex differentiation in an estrogen-like manner. The phytoestrogens and fungal mycotoxins provide some of the most conclusive data demonstrating that environmental estrogens are toxic to mammalian reproductive function under natural conditions.

Naturally occurring estrogenic compounds are widespread in nature. Farnsworth *et al.* (1975) listed over 400 species of plants that contain potentially estrogenic isoflavonoids or coumestans or were suspected of being estrogenic based on biological grounds. However, plants contain many other compounds in addition to estrogens that can affect reproductive performance. For example, it was also noted that of the 525 species of plants that have been used as folkloric contraceptives or interceptives, more than half of these were abortifacients that stimulated uterine muscle contractions *in vitro* (Salunkhe *et al.*, 1989). In addition, many plant extracts possess antispermatic agents that reduce sperm counts, some of which appear to act directly on the testis, or by altering hypothalamic–pituitary function (Salunkhe *et al.*, 1989). In recognition of the potency of these compounds and their potential to alter human reproduction, the World Health Organization of the UN has evolved common procedures for screening plant extracts for antifertility action.

Although most environmental estrogens are relatively inactive, as compared to steroidal estrogens or DES, they frequently occur in such high concentrations that they are able to have reproductive effects on animals (Adams, 1989). For example, soybean products may contain up to 0.25 % total isoflavones, concentrations which can cause signs of estrogenicity in swine. Soybeans and alfalfa sprouts can contribute detectable amounts of plant estrogens to the human diet (Adams, 1989). Pastures of clover containing formononetin cause fertility problems in sheep, having an estrogenic effect equal to 5–15 μg DES daily. Additive effects of these chemicals have been reported in domestic animals (Adams, 1989) when endogenous estrogen levels are low, as would be the case during sex differentiation of the male vertebrate, while, in contrast, they act antagonistically when steroid levels are high. Hence, it is possible that these chemicals are both additive and antagonistic in the same animal at different stages of the estrous cycle, or at different stages of development. Adams (1989) concluded that the amount of phytoestrogens are rarely high enough to totally disrupt fertility, so that subclinical effects are more common and, because phytoestrogens are so widespread, whole animal populations may be effected, either directly or via *in utero* exposure. For example, he concluded these chemicals cause at least 1 000 000 ewes to fail to lamb each year in Australia; an effect that is so widespread that this is accepted as 'normal' lambing rates (Adams, 1989). 'Clover disease', which is characterized by dystocia, prolapse of the uterus and infertility, is observed in sheep grazed on highly estrogenic clover pastures. Permanent infertility (defeminization) can be produced in ewes by much lower amounts of estrogen over a longer time period than are needed to produced 'clover disease'. Although there are no unequivocal reports of adverse effects associated with phytoestrogens in humans, during the Second World War II people in Holland consumed large quantities of tulip bulbs, containing high levels of estrogenic activity, and many women displayed signs of estrogenism, including uterine bleeding and other abnormalities of the menstrual cycle (Coussens and Sierens, 1949; Labov, 1977; Adams, 1989). In the laboratory, studies using rats have clearly demonstrated that plant estrogens can alter mammalian sex differentiation. For example, neonatal exposure to the plant estrogen coumestrol (at 100 µg per day for the first 5 days of life) defeminizes female offspring. Progeny display an accelerated age of vaginal opening, increased incidence of persistent vaginal cornification (PVC) and coumestrol-treated females had hemorrhagic follicles (100 % at 40 days of age) (Burroughs *et al.*, 1985).

### (b) Fungal toxin: zearalenone

In domestic animals, feeds contaminated with the zearalenone producing fungus *Fusarium sp.* induce adverse reproductive effects in a wide variety of domestic animals, including impaired fertility in cows and hyperestrogenism in swine and turkeys (Adams, 1989). In rodents, neonatal exposure to the fungal toxin zearalenone alters sex differentiation of the female rat and hamster reproductive system when administered as a single s.c. injection of 1.0 mg shortly after birth. Zearalenone causes PVC (Kumagai and Shimizu, 1982), and reduces size of the ovaries, due to a lack of corpora lutea, in rats while neonatally treated female hamsters display accelerated vaginal opening and abnormal male-like sex behavior as adults (Gray *et al.*, 1985). In contrast, similar treatment with the estrogenic mycotoxin zearalenone did not affect reproductive organ weights or male mating behavior in males (Gray *et al.*, 1985).

### (c) Pesticides and toxic substances

#### *(i) Alterations of sexual differentiation: effects in females*

It is known that perinatal administration of weakly estrogenic pesticides, like kepone (chlordecone), o,p′-DDT or methoxychlor, produces estrogen-like alterations on CNS sex differentiation in rodents. Weakly estrogenic chemicals accelerate vaginal opening, induce PVC and DAS (delayed anovulatory syndrome) in female rats, and masculinize sex-linked behaviors. Although apparently no longer a threat, occupational exposure to high levels of chlordecone resulted in reproductive and neurotoxicity in male workers, in a manner that closely resembles the effects of chlordecone in adult male rodents (Guzelian, 1982).

**o,p′-DDT** Administration of o,p′ DDT (1 mg, s.c.) on the second, third and fourth days of life to female rat pups induced PVC by 120 days of age; ovaries of the treated females contained follicular cysts and lacked corpora lutea (Heinrichs *et al.*, 1971). In the laboratory, mounting behavior and morphology are demasculinized in estrogen-treated (DES, estradiol or o, p′ DDT *in ovo*) male Japanese quail, while treated females display oviductal abnormalities and are unable to produce normal eggshells (Bryan *et al.*, 1989). In the field, pesticide exposure was correlated with abnormal reproductive tract development of the female California gull (Fry and Toone, 1981), an effect that was replicated in the laboratory by exposure to environmentally relevant concentrations of o, p′ DDT and methoxychlor.

**Chlordecone (kepone)** Although occupational exposure to chlordecone during the 'Kepone Incident' (Guzelian, 1982) altered reproductive function in male workers, fortunately there were no women (and hence, no human exposure during development) in the highly exposed population. In rodents, perinatal exposure to kepone produces effects typical for a weak estrogen (Gellert, 1978a). Injection of kepone (0.2 or 1.0 mg/pup) advanced vaginal opening by more than ten days and accelerated the onset of PVC in the high dosage group by 4 months of age and in the low dosage group at 6 months; ovarian weight was reduced in the high dose group due to the lack of corpora lutea. Administration of kepone to the dam during gestation (days 14–20, 15 mg $kg^{-1}$ $d^{-1}$) by gavage induced an early onset of PVC, anovulation and reduced ovarian weight and resulted in tonic serum levels of estradiol in 6-month-old female offspring (Gellert and Wilson, 1979). In the female hamster, neonatal injections of kepone or estradiol at 2 days of age masculinized the CNS such that treated females displayed high levels of male sex behavior as adults, a behavior rarely, if ever, seen in untreated females (Gray, 1982), In contrast, kepone-treated female hamsters are not defeminized, as they all displayed normal estrous cycles and feminine sex behavior despite being masculinized. Other studies using hamsters have shown that estrogen-induced

DAS, indicative of defeminization of the hypothalamus, is seen only at doses two orders of magnitude above those that masculinize females (Whitsett and Vandenbergh, 1975).

**Methoxychlor** Perinatal exposure to this weakly estrogenic pesticide produces the typical profile of reproductive alterations in female rat offspring. Oral administration of methoxychlor *in utero* and during lactation to the dam at 50 mg $kg^{-1}$ $d^{-1}$ induced PVC and the DAS in female offspring, and for this reason it reduced fecundity by 50 % compared to control females (Gray *et al.*, 1989).

**PCBs** A few of the 209 PCB congeners and PCB mixtures possess weak estrogenic activity. The Aroclor mixture Aroclor 1221 produces a uterotropic response in female rats and permanently alters neuroendocrine reproductive function in treated females, while Aroclor 1224 did not (Gellert, 1978b). Neonatal injection of 10 mg s.c. of Aroclor 1221 on the second and third days of life accelerated the age of vaginal opening, and increased the incidence of persistent vaginal estrus and anovulatory cycles. It has been proposed that these PCBs may be rendered estrogenic through metabolic hydroxylation, yielding polychlorinated hydroxybiphenyls (Korach *et al.*, 1987). When a series of PCB congeners were compared for estrogenic activity, 4-hydroxy-2′,4′,6′-trichlorobiphenyl bound to the estrogen receptor with the greatest affinity *in vitro*, and stimulated uterine weight increase *in vivo*. When two of these hydroxylated PCBs were administered *in ovo*, high doses altered sex differentiation in turtles such that an excess of females were produced (Bergeron *et al.*, 1994). These authors also found that coadministration of ineffective dosage levels of the two PCB metabolites produced synergistic alterations of reptilian sex differentiation.

### *(ii) Sexual differentiation: summary of effects in females*

Environmental estrogens produce a fairly uniform profile of abnormalities in female rodents after perinatal exposure, depending upon the species employed. Rats display irregular estrous cycles and prolonged estrus, which may precede the onset of PVC and DAS. In the female hamster, developmental exposure to estrogens accelerates vaginal opening but they are not easily defeminized and are fertile and continue to cycle even though they have been behaviorally masculinized. Adverse effects of synthetic environmental estrogens have also been reported in avian and reptilian species. In avian species, exposure to synthetic estrogens can alter gonadal sex differentiation and induce abnormal development of the right oviduct, while in reptiles lacking sex chromosomes, in which sex is determined by temperature, estrogenic PCBs can feminize differentiation of gonadal and reproductive tract, skewing the sex ratio in the female direction.

### *(iii) Alterations of sexual differentiation: effects in males*

Weakly estrogenic xenobiotics like o,p′DDT, chlordecone, zearalenone, alkylphenolic compounds and methoxychlor have the potential to alter sex differentiation of the fetal male in an estrogen- or DES-like manner. However, the effects that result from perinatal exposure to these substances are typically less severe in nature than those seen with DES.

**o,p′ DDT and methoxychlor** Gellert *et al.* (1974) found that neonatal male rats injected with methoxychlor or DDT had normal reproductive organ weights as adults. In addition, fertility was unaffected in methoxychlor-treated male offspring and epididymal cysts were not detected (Gellert and Wilson, 1979). However, when methoxychlor was administered to the dam throughout gestation and lactation subtle alterations were noted in the treated males, such as slightly smaller testes, epididymides and lower sperm counts compared to controls (Gray *et al.*, 1989).

It is noteworthy that adverse reproductive development has been reported in the field in avian species that are not susceptible to eggshell thinning after *in ovo* exposure to environmental levels of o,p′DDT or methoxychlor. Pesticide treatment feminized the behavior, the gonads and reproductive tracts of the male California gull (Fry and Toone, 1981) and the decline in breeding success and local population levels of this species were attributed to these xenobiotic-induced reproductive alterations. In the laboratory, mounting behavior and morphology are demasculinized in estrogen-treated (DES, estradiol or o, p′ DDT) male Japanese quail (Bryan *et al.*, 1989).

**Chlordecone (kepone)** Kepone produces a similar pattern of reproductive alterations in adult men and rodents. Men exposed to kepone in the workplace displayed reduced libido and reduced sperm counts (Guzelian, 1982). Although there are no data on the effects of this pesticide on human sexual differentiation, in hamsters, partial demasculinization of the male reproductive resulted from a single injection of chlordecone. Treated hamsters had reduced testicular and epididymal weights, but male sexual behavior was unaffected and the epididymides did not develop granulomas (Gray, 1982).

### *(iv) Alterations of pubertal development*

In addition to the effects of estrogenic chemicals on sex differentiation, subchronic exposure of weanling male and female rats to environmental estrogens can result in alterations of pubertal development. Exposure to a weakly estrogenic pesticide after weaning and through puberty induces pseudoprecocious puberty (accelerated vaginal opening without effect on the onset of estrous cyclicity) in the female, but causes antiandrogen-like delays in males. In humans, exposure to estrogen containing creams or drugs can induce pseudoprecocious puberty in young girls. An outbreak of pseudoprecocious puberty was reported in Puerto Rico with 482 cases occurring in girls between the ages of 6 and 24 months (Hannon *et al.*, 1987) between 1976 and 1984. However, the cause of this apparent outbreak was never determined.

**Effects in females** Initiation of subchronic treatment with methoxychlor at weaning (25 mg $kg^{-1}$ $d^{-1}$ or higher) causes pseudoprecocious puberty in female rats. Vaginal opening occurs from 2 to 7 days earlier, in a dose-related fashion, but methoxychlor treatment does not accelerate the onset of regular estrous cycles, indicating a direct estrogenic effect of methoxychlor on vaginal epithelial cell function without an effect on hypothalamic-pituitary maturation. Similar effects have been achieved with chlordecone, another weakly estrogenic pesticide, and octylphenol; however, chlordecone also induces neurotoxic effects (hyperreactivity to handling). Although methoxychlor induces female mating behavior in an estrogen-like manner, both methoxychlor (Gray *et al.*, 1988a) and the phytoestrogens (Adams, 1989) seem relatively ineffectual in altering pituitary function in either male or female rats.

**Effects in males** Subchronic methoxychlor administration delays the onset of puberty (preputial separation) in male rats at doses of 100–400 mg $kg^{-1}$ $d^{-1}$ in a dose-related manner. In some cases, puberty was delayed by as much as 10 weeks and time to pregnancy, later in life, was delayed. In addition, methoxychlor-treated males have smaller seminal vesicles and ventral prostates, lower epididymal and ejaculated sperm numbers and copulatory plug formation is inhibited (Gray *et al.*, 1989; Anderson *et al.*, 1995). It is possible that the antiandrogenic effects of methoxychlor are mediated via an active metabolite (HPTE, Table 4.2) that competitively inhibits androgen receptor binding. In fact, most of the synthetic estrogenic chemicals that we have studied to date also possess antiandrogenic activity although this may not be the case for the phytoestrogens (Laws *et al.*, 1995). In contrast, the effects of methoxychlor on the female and the inhibition of food consumption in the male rat, seen at 25 mg $kg^{-1}$ $d^{-1}$, are characteristic of estrogenic rather than antiandrogenic activity.

### *4.4.3 ENVIRONMENTAL ESTROGENS: FUTURE DIRECTIONS*

As new classes of synthetic environmental estrogens emerge, their potential to alter mammalian, avian and reptilian sex differentiation must be evaluated. For example, exposure to phenols and alkylphenol ethoxylates (APEs) in rivers in England results in vitellogenesis, an estrogenic-response, gonadal atrophy and hermaphrodism in male fish (Purdom *et al.*, 1994). We have found that two of these phenols also bind to androgen as well as estrogen receptors (Laws *et al.*, 1995) *in vitro*. The potential health effects of this class of environmental estrogens need to be assessed because millions of tons of these chemicals are used in the production of plastics and they can occur

**Table 4.2** Relative binding affinity (RBA) of select steroids, pharmaceuticals and environmental chemicals to estrogen (ER), androgen (AR) and progesterone (PR) receptors.

| Chemical | RBA to ER | RBA to AR | RBA to PR |
|---|---|---|---|
| *RBA of steroids for the estrogen, androgen and progesterone receptors* | | | |
| Estradiol[a] | +++ | ++ | 0 |
| Testosterone | + | +++ | + |
| DHT[a] | + | +++ | + |
| Progesterone[a] | 0 | 0/+ | +++ |
| Cortisol[a] | 0 | 0 | 0 |
| Corticosterone[a] | 0 | 0/+ | + |
| Aldosterone[a] | 0 | 0 | 0/+ |
| *RBA of select pharmaceuticals to the estrogen, androgen and progesterone receptors* | | | |
| Levonorgestrel[b] | 0 | ++ | +++ |
| Norethidrone[b] | ++ | + | +++ |
| Norethidrone acetate[a] | ND | 0 | + |
| Norgestrel[a] | ND | +/++ | +++ |
| Norgestrienone[a] | ND | + | + |
| Promegestrone[a] | ND | 0/+ | +++ |
| Demegesterone[a] | ND | 0/+ | +++ |
| Chlormadinone acetate[a] | ND | + | +++ |
| Gestrinone[a] | ND | + | + |
| Triamcinolone acetonide[a] | 0 | 0 | + |
| Dexamethasone[a] | 0 | 0 | 0/+ |
| RU1881[a] | 0 | +++ | +++ |
| RU26988[a] | 0 | 0 | 0 |
| MPA[b] | 0 | + | +++ |
| Megestrol acetate[b] | 0 | 0 | +++ |
| DES[c,d] | +++ | ++ | ND |
| *RBA of pesticides and toxic substances to estrogen, androgen and progesterone receptors* | | | |
| Chlordecone[c,d] | ++ | + | ++ |
| o,p′-DDT[c,d] | ++ | + | ++ |
| p,p′-DDT[c,d] | 0 | + | 0 |
| p,p′-DDE[c,d] | 0 | ++ | 0 |
| p,p′-DDD[c,d] | 0 | + | 0 |
| Lindane [c,d] | 0 | ND | ND |
| Vinclozolin (V)[c,d] | 0 | 0 | 0 |
| V metabolite M1[c,d] | 0 | + | 0 |
| V metabolite M2[c,d] | 0 | ++ | + |
| Methoxychlor[d] | 0 | 0 | 0/+ |
| HPTE[d] | ++ | ++ | ++ |
| Zearalenone[e] | +++ | + | ND |
| Coumesterol[e] | +++ | 0 | ND |
| Genistein[e] | +++ | 0 | ND |
| Endosulfan[c,d] | 0 | ND | ND |

**Table 4.2** Continued.

| Chemical | RBA to ER | RBA to AR | RBA to PR |
|---|---|---|---|
| Atrazine[c,d] | 0 | ND | ND |
| Octylphenol[c,d] | ++ | + | ND |
| Nonylphenol[c,d] | ++ | + | ND |

KEY: +++ = very active; ++ = active; + = weakly active; 0 = inactive; ND = no data.
[a] Vannier and Raynaud, 1980; [b] Stanczyk, 1994; [c] Kelce *et al.*, 1995; Kelce *et al.*, 1994; [d] Laws *et al.*, 1995; [e] Kelce and Laws, unpublished data.

locally at high concentrations in the environment and at low concentrations in some food liners.

Our understanding of the toxicity of synthetic environmental estrogens will be enhanced if we recognize that these chemicals typically bind to other steroid hormone receptors, often with affinity equivalent to that displayed for the estrogen receptor. Comments that the estrogen receptor is 'sloppy' but other receptors are not ignore recent data indicating that the androgen receptor also is promiscuous. As indicated earlier, all of the synthetic environmental estrogens that we have studied to date display measurable binding affinity for the androgen receptor. Some of the 'estrogens' like nonylphenol and the methoxychlor metabolite HPTE (Table 4.2) bind to the androgen and/or the progesterone receptors with about the same affinity displayed for the estrogen receptor.

Greater care should be taken to reduce the amount of misinformation about environmental estrogens communicated by the press and in scientific journals. Numerous toxicants have been described as possessing estrogenic activity based on incomplete or inaccurate analyses, while in other cases, the scientific data are ignored. For example, non-estrogenic chemicals (p,p′DDE) are repeatedly reported to be estrogenic, in spite of the fact that p,p′DDE and p,p′DDT, the primary p,p′DDT metabolites found in human and animal tissues, do not bind to the estrogen receptor in cell-free receptor assays at toxicologically relevant concentrations, and they are not estrogenic *in vivo*. It makes little sense to look for estrogenic effects in humans or wildlife from exposure to p,p′DDT and p,p′DDE.

We should also be careful about what we accept as proof of endocrine/estrogenic activity. Concerns have recently been directed at the potential endocrine effects of a number of pesticides and toxic substances, including endosulfan (which is widely used and found in food) and atrazine (which is also widely used and found in drinking water), which require verification of estrogenicity in *in vivo* and *in vitro* test systems, as these pesticides have not yet been studied in a comprehensive manner. Although the pesticide endosulfan has been reported to be estrogenic, based on MCF7 cell proliferation *in vitro*, endosulfan fails to bind to the estrogen receptor *in vitro* (Laws *et al.*, 1995). It is possible that endosulfan is metabolized to an estrogen, and is estrogenic *in vivo*, but without additional data on this chemical it is too early to conclude that endosulfan is estrogenic *in vivo*. In addition, some of the chemicals listed as 'endocrine disrupters' in Table 4.1 are not only not estrogenic, they are not even direct-acting endocrine toxicants. Some of these chemicals only affect the endocrine system indirectly.

## 4.5 MECHANISMS OF ENDOCRINE TOXICITY: ENVIRONMENTAL ANTIANDROGENS

### *4.5.1 INTRODUCTION*

Although many reproductive toxicants bind to the estrogen receptor and act as estrogen agonists or antagonists the focus of the study of 'endocrine-disrupting' toxicants needs to be expanded beyond estrogenicity. Consideration (research and funding) should also be given to other mechanisms of endocrine toxicity – mechanisms that, in fact, may be of equal or greater concern. Although there are hundreds of naturally occurring plant and fungal estrogens which adversely affect reproductive function in domestic animals and there are a large number of synthetic estrogenic toxicants, there are also large numbers of naturally occurring and synthetic toxicants that act as 'endocrine disrupters' through alternative mechanisms.

Our studies have revealed that synthetic environmental antiandrogens are another important group of endocrine disrupters. In addition, there are naturally occurring phytoantiandrogens (Carilla *et al.*, 1984). Permixon is a drug used in the treatment of BPH extracted from the fruit of the dwarf palm tree (*Serenoa repens* B, native to Florida). Apparently, the antiandrogenic effect of this natural antihormone is mediated directly via dual action at the level of the androgen receptor and via inhibition of 5α-reductase activity.

One anthropocentric environmental androgen has been described. Microbially transformed phytosteroids present in streams receiving kraft mill effluent masculinized the morphology and behavior of female live-bearing Poeciliidae fish in an androgen-like manner (Davis and Bortone, 1992). These field observations were replicated with laboratory exposures to microbially treated phytosterol; however, the specific substance in effluents remains unidentified.

### *4.5.2 ANTIANDROGENIC DRUGS: MODELS FOR ENVIRONMENTAL ANTIANDROGENS*

Abnormal sexual development in man and rodents can be induced by drugs with antiandrogen activity (Schardein, 1993). Androgen receptor antagonists or inhibitors of DHT synthesis cause profound alterations in T- and DHT-dependent phenotypic sex differentiation (Wilson, 1992). However, in rodents the pattern of malformations produced by flutamide, which inhibits T and DHT binding to the intracellular androgen receptor, differs considerably from that produced by inhibiting only DHT synthesis with finasteride, a 5α-reductase inhibitor (Imperato-McGinley *et al.*, 1992). Specifically, both antiandrogenic agents flutamide and finasteride demasculinize external genitalia of male offspring, while only the receptor antagonist flutamide induces a high incidence of undescended ectopic scrota/testes, very small to absent prostates and inhibits T-mediated Wolffian duct differentiation.

### *4.5.3 AN ANTIANDROGENIC FUNGICIDE: VINCLOZOLIN*

The fungicide vinclozolin, which fails to bind to the estrogen receptor, alters male rat sex differentiation in a flutamide-like manner because it competitively binds to and inhibits the action of androgen receptors (AR) (Kelce *et al.*, 1994). Perinatal exposure to vinclozolin induces hypospadias, ectopic testes, a vaginal pouch, agenesis of the ventral prostate and nipple retention in male rat offspring (Gray *et al.*, 1994). In contrast, the female offspring are unaffected (Gray *et al.*, 1994). The developing fetus is very sensitive to endocrine disrupters like vinclozolin. Infertility, hypospadias and reduced ejaculated sperm counts are seen in male offspring whose dams were exposed to vinclozolin during sex differentiation (gestational day 14 to postnatal day 3) at 50 mg $kg^{-1}$ $d^{-1}$ and above and reductions in ventral prostate weight were seen at doses as low as 6–12 mg $kg^{-1}$ $d^{-1}$ and above. In contrast, even long-term high dose (100 mg $kg^{-1}$ $d^{-1}$ for 25 weeks) exposure does not reduce fertility in adult male rats (Fail *et al.*, 1995).

Weanling male rats exposed to vinclozolin through young adulthood exhibit delayed puberty (as indicated by preputial separation and the weight at puberty) and a number of endocrine and morphological effects; effects that are diagnostic of an antiandrogen (Monosson *et al.*, in preparation). Pubertal vinclozolin administration reduces seminal vesicle, ventral prostate and epididymal weight because vinclozolin metabolites block the stimulatory effect of androgens on these tissues. Serum LH is increased at low doses, because the negative feedback of androgens is blocked. The elevation in LH results in an elevation of testosterone levels in the serum and testosterone production by the testes *in vitro*. An examination of AR indicated that vinclozolin administration altered the distribution of AR, lowering the percentage of ARs extractable in a high-salt buffer (i.e. bound to DNA) without reducing the total numbers of AR. Procymidone, a fungicide from the same class as vinclozolin, produces similar developmental abnormalities in male offspring after perinatal exposure.

### *4.5.4 ANOTHER ENVIRONMENTAL ANTIANDROGENIC: p,p′DDE*

In recent work (Kelce *et al.*, 1995) we demonstrated, both *in vivo* and *in vitro*, that p,p′DDE is another potent environmental antiandrogen. *In vivo*, p,p′DDE delays puberty to the same degree as does vinclozolin. In addition, maternal exposure during sex differentiation reduces anogenital distance and induces retained nipples in male progeny. These *in vivo* effects are not mediated by stimulation of sterol hydroxylases and enhanced testosterone turnover because serum testosterone was not reduced by p,p′DDE administration in any of our *in vivo* studies. In addition, castrate plus testosterone-treated adult male rats have reduced ventral prostate and seminal vesicle weights in spite of the fact that serum testosterone levels are not reduced. *In vitro*, p,p′DDE binds to the androgen receptor and prevents DHT-induced transcriptional activation in cells transfected with the human androgen receptor. Further *in vivo* and *in vitro* laboratory studies of the antiandrogenic effects of p,p′DDE are warranted because DDT is still used in many areas of the world and the concentrations of DDE that are effective *in vitro* are lower than some human tissue and milk levels in these areas. In these countries, it is of particular concern that the amounts of p,p′DDE in human breast milk and dairy products not only exceed WHO standards, but they also exceed the *in vitro* LOEL for DDE. The fact that the *in vitro* LOEL is exceeded is of concern because with vinclozolin we found that 100% of the male offspring displayed hypospadias and other malformations when the concentration of vinclozolin metabolites M1 and M2 in maternal serum approached the $K_i$ (Kelce *et al.*, 1994).

It is important to note that, in addition to acting as an antiandrogen, p,p′ DDE produces a number of endocrine alterations in mammalian and wildlife species that result from other mechanisms of toxicity. For example, p,p′ DDE and o,p′ DDD alter adrenal physiology directly and indirectly, which inhibits steroid hormone synthesis. These DDT metabolites also act as endocrine disrupters by reducing hormone levels via increased catabolism of steroid hormones in both rodents and humans (Poland *et al.*, 1970). In addition, p,p′ DDE levels of about 2–14 ppm are associated with reproductive failure in raptors (Spitzer *et al.*, 1978) due to eggshell thinning; an effect seen in natural populations that has been reproduced in laboratory studies. p,p′ DDT/DDE exposure inhibits Ca-ATPase, carbonic anhydrase and calcium levels in the oviduct of susceptible avian species. Ring doves given 10 ppm p,p′DDT showed increased hepatic activity and a decrease in serum estradiol, leading to a delay in egg laying, a decrease in bone calcium deposition and eggshell weight (Peakall, 1970). In the field, DDT exposure has been linked to developmental reproductive abnormalities (small penis, abnormal hormone levels, skewed sex ratio with an increase in the percentage intersex and decrease in percentage male) in alligators in Lake Apopka (Guillette *et al.*, 1994). Alligator eggs from Lake Apopka contained levels of DDE (5.8 ppm) that are well above the concentrations that block androgen receptor function *in vitro*. Among the multiple mechanisms by which DDE acts as an endocrine disrupter we suspect that the antiandrogenic activity of DDE plays a role in the alterations in sex differentiation seen in both avian (Fry and Toone, 1981) and reptilian species (Guillette *et al.*, 1994) because the levels of DDE seen in these wildlife populations greatly exceed the concentrations of DDE that are antiandrogenic *in vitro*.

Although human/wildlife exposure to high levels of these chemicals is typically less than 20–40 years ago, high levels of these chemicals still persist in some areas of the North American continent. This fact is not widely recognized and it is generally assumed that problems with DDT and its metabolites are behind us because industrial production of most bioconcentrating chemicals was stopped in the USA during the 1970s and their presence in the environment has diminished. However, such conclusions ignore the fact that high levels of DDT metabolites, levels that often exceed WHO guidelines by more than ten-fold, are found in human milk in highly exposed populations in India (Battu *et al.*, 1989), Turkey and Canada (Dewailly *et al.*, 1989). Even though application of DDT was ended in the 1970s, extremely high environmental levels of DDT metabolites are still found in several avian populations. American robin eggs from orchard areas in British Columbia in 1991 contained extremely high levels of both DDE and DDT (total burden of 130 mg $kg^{-1}$); levels high enough to conclude that a considerable hazard existed to birds of prey because some raptors begin to display eggshell thinning at 2 mg $kg^{-1}$ (Elliot *et al.*, 1994). High levels still persist in the field because some orchards received more than 1000 kg of DDT/ha from the late 1940s through 1960. High levels of DDE are also found in human milk as a consequence of consumption of contaminated food, including marine mammals (Dewailly *et al.*, 1989), milk and dairy products (Battu *et al.*, 1989; Bouman *et al.*, 1990). In fact, these scientists determined that daily intakes of DDT and its metabolites through consumption of contaminated milk by 1–3-year-old children exceeded their acceptable daily intake by three to five fold (Battu *et al.*, 1989) and that a well-founded risk to firstborn infants exists.

### *4.5.5 ANTIANDROGENIC ACTIVITY OF ESTROGENS: DES, ESTRADIOL, CHLORDECONE, PHENOLS AND o,p′DDT*

Although environmental antiandrogens are just beginning to receive attention, it is clear that many other chemicals may display this antihormonal activity. For example, we have found that synthetic estrogenic chemicals, including estradiol, DES, chlordecone, nonylphenol, octylphenol and o,p′ DDT, all bind to the androgen receptor (Table 4.2; Laws *et al.*, 1995). For those chemicals studied to date, effects seen in cells transfected with human androgen receptors containing the luciferase reporter gene demonstrate that these chemicals act at low concentrations as androgen receptor antagonists rather than agonists. These results suggest that the demasculinizing action of estrogenic substances on the male offspring after *in utero* administration, or the pubertal delays and sex accessory weight reductions seen in male rats exposed to estrogens, may result from an antagonistic interaction of the 'estrogenic' compound with the androgen receptor, rather than acting as an estrogen agonist (Table 4.2).

Interestingly, some environmental estrogens, like chlordecone and o,p′ DDT, also bind to the progesterone receptor *in vitro* with about the same affinity that they display for the estrogen receptor (Laws *et al.*, 1995). The fact that environmental endocrine disrupters can bind to multiple steroid hormone receptors should not be surprising because similar effects have been extensively reported both for steroidogenic drugs and for the steroid hormones themselves (see Table 4.2). While some of these toxicants also inhibit the enzyme 5α -reductase that converts testosterone to the more potent androgen DHT, this effect often occurs at higher *in vitro* concentrations (Laws *et al.*, 1995).

## 4.6 OTHER MECHANISMS OF ENDOCRINE TOXICITY

### *4.6.1 EFFECTS OF TCDD ON REPRODUCTIVE DEVELOPMENT*

2,3,7,8 Tetrachlorodibenzo-p-dioxin (TCDD) is an ‘endocrine disrupter’ that acts on more than one component of the endocrine axis at a time (Birnbaum, 1994). For this reason, it is much too simplistic to continue to view this ubiquitous toxicant only as an antiestrogen, as some scientists have proposed. TCDD exposure alters the levels of many hormones, growth factors and their receptors and hormone synthesis through interaction with the Ah receptor. During development, administration of low doses of TCDD or PCB 169 (a dioxin-like congener) alter reproductive development and fertility of the progeny, demonstrating that the process of sexual differentiation in the developing rodent fetus is extremely sensitive to 2,3,7,8 TCDD and related chemicals. For example, the offspring of Wistar rats dosed with 0.5 μg $kg^{-1}d^{-1}$from gestational day (GD) 6 to 15 display reduced fertility (Khera and Ruddick, 1973). In a three generation reproduction study (Murray *et al.*, 1979), TCDD reduced fertility of Sprague–Dawley rats in the F1 and F2 but not the F0 (no developmental exposure) generation at 0.01 μg $kg^{-1}$ $d^{-1}$ in the diet. Administration of a single dose of TCDD to the dam on GD 15 at doses between 0.05 to 1 μg demasculinizes Long Evans (LE) Hooded male rat offspring (LOAEL = 0.2) (Gray *et al.*, 1995a; Gray and Ostby, 1995). In addition, we found that GD 11 treatment, equivalent to GD 15 in the rat, at 2 μg/kg, demasculinized male Syrian hamster offspring (Gray *et al.*, 1995a). In LE rat and hamster male offspring, puberty (preputial separation) was delayed by about 3 days, ejaculated sperm counts were reduced by at least 58% and epididymal sperm storage was reduced by 38%. Testicular sperm production was less affected. The sex accessory glands also were reduced in size in LE rat offspring treated on GD 15 in spite of the fact that serum testosterone (T), T production by the testis *in vitro*, and androgen receptor (AR) levels were not reduced. Some reproductive measures, such as anogenital distance and male sex behavior, were altered by TCDD treatment in rat but not hamster offspring. Since T and AR levels were not reduced in sex accessory glands and the epididymis following perinatal TCDD exposure, the alterations in these tissues are not likely to have resulted from an alteration of the androgenic status of the male offspring. These observations replicate some, but not all of the effects of TCDD on male Holtzman rat progeny, reported by Mably *et al.* (1992 a,b,c).

We also discovered that TCDD-treated female rats and hamsters also displayed a number of unusual reproductive alterations (Gray and Ostby, 1995; Gray *et al.*, 1995b, LOAEL = 0.2 μg $kg^{-1}$). Many of the rat progeny displayed clefting of the phallus with mild degree of hypospadias. Most female progeny from the GD 15 TCDD treatment group displayed a permanent 'thread' of tissue across the opening of the vagina. Similar morphological alterations were noted in TCDD-exposed Holtzman rats. When TCDD-treated LE female offspring were mated to untreated males on the day of vaginal proestrus, female sexual behaviors (darting and lordosis to mount ratios) were normal; however, males had difficulty attaining intromission and vaginal bleeding was displayed by most TCDD-exposed female offspring during mating. At the end of their reproductive life, GD 8 TCDD-treated female progeny displayed reduced levels of fertility, a higher incidence of constant estrus and cystic endometrial hyperplasia. Female hamsters displayed clitorine clefting and reduced fertility, but not the vaginal 'thread'. Taken together, administration of a single dose of TCDD late in gestation results in similar malformations of the external genitalia in two species of rodents that differ greatly in their sensitivity to the lethal effect of TCDD.

### *4.6.2 EFFECTS OF PCBS AND PCDFS ON REPRODUCTIVE DEVELOPMENT*

Exposure to polychlorinated biphenyls (PCBs) and polychlorinated dibenzofurans (PCDFs) is clearly linked to developmental/reproductive toxicity in humans, primates, rodents, mink, fish and other wildlife species. A recent review of the reproductive toxicity of commercial PCB mixtures from animal studies (Golub *et al.*, 1991), identified LOAELs of 0.1–0.2 mg $kg^{-1}$ $d^{-1}$ for reproductive effects in mink rats and mice. Non-human primates appeared to be affected by PCB toxicity at lower doses than rodent species (Golub *et al.*, 1991). In addition to the reproductive effects seen in animal studies, these chemicals dramatically alter human development. *In utero* exposure to a mixture of PCB/PCDFs in 1978–79 in Taiwan resulted in 'Yu-Cheng' illness in children characterized by hyperpigmentation, nail deformities, and poorer cognitive and behavioral development. The physical and cognitive problems are known to persist for at least ten years (Lai *et al.*, 1993). Recently, it was noted that exposed boys with 'Yu Cheng' illness had a shorter penis at 11–14 years of age in addition to the aforementioned physical and CNS alterations (Guo *et al.*, 1993).

In other populations, exposure to PCBs can occur at high levels. For example, a study of transformer maintenance workers, exposed primarily to Aroclor 1260 for four years, reported a significant decrease in serum T4 levels (Emmett *et al.*, 1988). McKinney (1987) calculated that background levels of

PCBs in the milk may be sufficiently high to cause neonatal hypothyroidism in exposed babies, an effect that could be responsible for the effects of PCBs in children. In the US children exposed to PCBs at levels considered to be background have hypotonia, hyporeflexia, delayed psychomotor development at 6 and 12 months, and poorer visual recognition memory at 7 months (Tilson *et al.*, 1990). For this reason, it is of particular concern that exposure to high levels of PCBs in human milk has been documented. Doses of PCBs as high as 62 μg $kg^{-1}$ $d^{-1}$ or 514 μg $l^{-1}$ of milk, with an average of 13 μg $kg^{-1}$ $d^{-1}$ were detected from women in Northern Quebec (Dewailly *et al.*, 1989). This 'average' is 13-fold above the Acceptable Daily Intake of 1 μg $kg^{-1}$ (Health and Welfare Canada). In fact, Dewailly *et al.* (1989) calculated that a baby exposed to 0.5 mg PCBs $l^{-1}$ of milk would attain a blood level of 150 μg $kg^{-1}$ in 3 months (the human LOAEL; NIOSH, 1977). This blood level is set as the LOAEL because it produces clinical and biological effects (chloracne, hepatic, enzymatic induction) in the most sensitive workers.

The fact that TCDD produces alterations of rodent sex differentiation immediately causes one to speculate about the potential for other dioxin-like chemicals, including polychlorinated dibenzofurans (PCDFs) and polychlorinated biphenols (PCBs), to induce similar alterations of reproductive development. In fact, when TCDD-like PCB congener 169 is administered to the dam on gestational day 1 or shortly after implantation at 1.8 mg $kg^{-1}$, reduced fertility was seen in both male and female offspring (Smits-van Prooije *et al.*, 1993; Gray, unpublished). The reproductive system of the developing animal appears to be more sensitive to PCBs than during adulthood, with adult males being the least sensitive. During lactation, exposure to Aroclor 1254 (dams dosed at 32 mg $kg^{-1}$ $d^{-1}$) reduces the reproductive capacity of male and female offspring (Sager, 1983; Sager *et al.*, 1987; 1991; Sager and Girard, 1994). Similar doses reduce the reproductive potential of the adult female (Linder *et al.*, 1974; Jonsson *et al.*, 1976; Orberg and Ingvast, 1976; Brezner *et al.*, 1984). In contrast, such doses have little effect on the reproductive system of the adult male rat (Gray *et al.*, 1993). The increased testis size in male rats treated with PCB 1254 during lactation (Sager, 1983), or Arochlor 1242 during infantile stage of life (Hess and Cooke, 1995) probably results from PCB-induced hypothyroidism during a critical stage of testicular development. In the rat, these effects closely resemble the effects of PTU-induced hypothyroidism on testicular development and function (Hess and Cooke, 1995). PCB and PTU-treated male rats develop testes that are considerably larger and produce many more sperm than normal in spite of their reduced body size. Presumably the lack of thyroid hormone during development prolongs the period of Sertoli differentiation in the testis.

In addition to altering mammalian reproduction, PCBs induce well-documented alterations of reproductive function in lower vertebrates as well. PCBs and TCDD are found in high concentrations in a number of lakes and estuaries. These chemicals are readily bioaccumulated by fish, and although

**Table 4.3** Summary of reproductive effects of laboratory exposure of adult or juvenile fish to PCBs

| Species | Dose ($mg\ kg^{-1}$) (ppm) | Tissue conc. ($mg\ kg^{-1}$) | Effects (ppm) | Reference |
|---|---|---|---|---|
| Carp | 25 × 4 (i.p.) over 4 wk A1254 | ND[a] | ↓ Androgen<br>↓ Estrogen<br>↓ Corticoids | Sirvarajah, *et al.*, 1978 |
| Trout | 25 × 4 (i.p.) over 4 wk A1254 | ND | ↓ Androgen<br>↓ Estrogen<br>↓ Corticoids | Sirvarajah, *et al., 1978* |
| Atlantic croaker | 3.4 in diet for 30 d A1254 | ND | ↓ Ovarian growth by 75%<br>↓ Testosterone | Thomas, 1988 |
| Atlantic croaker | 5 in diet for 17 d A1254 | ND | ↓ Ovarian growth (GSI) by 50%<br>↓ Estradiol<br>↓ Gonadotropin from pit | Thomas, 1989 |
| Atlantic croaker | 1 in diet for 30 d A1254 (adults) | ND | ↑ % Abnormal embryos<br>↓ % Hatch<br>↓ % Viable hatch<br>↓ larval length | Sharp and Thomas, 1991 |
| Atlantic cod | 1,10,25, 50 in diet for 5 mth A1254 | 0:2–9<br>1:7–14<br>10:45–100<br>25:156<br>50:374<br>(liver, wet) | Altered steroid prod.<br>Abnorm. testes<br>↓ Spermatogenic elements | Freeman *et al.*, 1982 |
| Brook trout | 3.8 ppm in water for 21 d A1254 | 77.9 in eggs<br>32.8 in skeletal muscle | ↓ Testes size<br>↓ Spermatic fluid<br>↓ % hatch | Freeman and Idler, 1975 |
| Minnow | 20,200, 2000 in diet 40 d A1254 | 20:1–10<br>200:20–30<br><br><br>(whole body, wet) | 200: spawn delayed 7 d<br>2000: spawn delayed 21<br>2000:100–300<br>Mortality at high dose | Bengtsson, 1980 |
| Fundulus | 1,10,100 (i.p.) 3,3′,4,4′-TCB | ND | ↓ Egg deposition by:<br>1:49%<br>10:78%<br>100:96%<br>50% mortality at 100, 25% at 10 | Black and McElroy, 1991 |
| White perch | 0.2, 1, 5 (i.p.) 3 × over 3 mth 3,3′,4,4′-TCB | 0.2:0.4–0.8<br>1.0:2.4–4.0<br>5.0:3.7–4.5<br>(liver, wet) | At 5:<br>↓ % mature females<br>↓ GSI (male and female)<br>At 5 and 1<br>↓ % larval survival | Monosson *et al.*, 1994 |

[a]No data.

**Table 4.4** Concentrations of PCBs in aquatic animals

| Species | PCB total (ppm) | 3,3′,4,4′-TCB (ppb) | Ref. |
|---|---|---|---|
| Striped mullet<br>Seto Island, Japan | 0.22–3.2<br>muscle | 0.58–4.8<br>muscle | Tanabe *et al.*, 1987 |
| Killer whale<br>Japan | 160<br>blubber | 42<br>blubber | Tanabe *et al.*, 1987 |
| Pacific white-sided dolphin<br>Iko Island, Japan | 14–38<br>blubber | 40–71<br>blubber | Tanabe *et al.*, 1987 |
| Striped bass<br>Hudson River<br>NY Bight | 3.1–30.3<br>muscle | 9–68.3<br>muscle | Hong and Bush, 1990 |
| Baltic salmon<br>Baltic Sea | 0.33<br>fillet | 0.64<br>fillet | Tarhanan *et al.*, 1989 |
| Black bullhead | 49 (body) | 89 (body) | Huckins *et al.*, 1988 |
| White sucker | 41 (body) | 50 (body) | |
| Yellow perch<br>Waukegan Harbor, Lake Michigan | 11 (body) | 23 (body) | |
| Winter flounder | liver, dry weight: | liver, dry weight | |
| New Bedford H. | 44–380 | 40–1150 | Elskus, 1992 |
| Narragansett | 1–27 | n.d.–10 | Elksus, 1992 |
| Georges Bank | 0.3–0.8 | n.d.–3 | Monosson and Stegeman 1994 |

half-lives for congeners vary (Niimi and Oliver, 1983), some tend to be much longer in fish than in mammals (e.g. 3,3′,4,4′-TCB, a planar PCB congener, has a half-life of only 3 days in the adipose tissue of rats (Clevenger *et al.*, 1989), while it can be up to 44 days in whole fish (Niimi and Oliver, 1983)). For these reasons, the tissues of fish in many areas exceeds federal standards for human consumption, and can be an important source for human exposure to PCBs and similar compounds.

PCBs adversely affect the hypothalamic–pituitary-gonadal axis in adult fish (Tables 4.3 and 4.4) resulting in reduced gonadal growth, reduced steroid hormone and gonadotropin concentrations and reduced egg deposition (see Table 4.3). Developing larvae can be exposed to lipophilic compounds such as PCBs through the egg yolk. In fact, body burdens of PCBs in female fish following spawning may be lower than before egg production began (Vodicknik and Peterson, 1985) and reduced larval survival following maternal exposure to PCBs has also been demonstrated (Monosson *et al.*, 1994; Sharp and Thomas, 1991).

Environmental exposure to PCBs has also been associated with reproductive impairment in natural populations of marine and freshwater fish. Similar to the laboratory studies, suppressed ovarian growth, and lowered levels of serum steroid hormones (Casillas *et al.*, 1991; Johnson *et al.*, 1988) have all been associated with exposures to PCBs. It is important to note that concurrent exposure to mixtures of heavy metals and polyaromatic compounds is common at many study sites. Decreased embryo and larval survival has also been associated with maternal exposure to PCBs and other anthropogenic contaminants in field studies (Spies and Rice, 1988; von Westernhagen *et al.*, 1981).

More recent studies with either the planar dioxin-like PCB congeners or with dioxin show that some of the effects observed with PCB mixtures may in part be caused by PCBs binding to the Ah receptor such as 3,3′,4,4′-tetrachlorobiphenyl (Monosson *et al.*, 1994; Walker and Peterson, 1992; Wannemacher *et al.*, 1992). Together these studies demonstrate that PCBs produce alterations of reproductive function in fish at several levels of the reproductive system.

The PCB/PCDFs provide an extreme example of the complex/multiple mechanisms by which toxicants act on the endocrine system. While a few of these chemicals possess TCDD-like activity, which in itself disrupts several components of the endocrine system, PCBs and PCDFs congeners that do not bind to the Ah receptor can also alter the hormonal milieu via Ah-independent mechanisms. Some of the additional mechanisms of action that may be displayed by an individual chemical within these classes include effects on thyroid hormone activity, neuroendocrine alterations, induction of liver metabolism of steroid and other hormones, and inhibition of hormone synthesis. For example, it has been hypothesized that PCBs reduce serum hormones, like T4, in part by inducing hepatic microsomal enzyme activity by increasing biotransformation, deactivation and biliary excretion of T4 (Bartomsky, 1974; Bartomsky *et al.*, 1976). Aroclor 1254 greatly increases hepatic UDP-glucuronosyltransferase activity, leading to an increase in UDP-GT activity towards T4, leading to a reduction in both total and free T4 levels (Barter and Klaassen, 1992).

## 4.7 POTENCY OF ENVIRONMENTAL ENDOCRINE DISRUPTERS

It is often reported that environmental endocrine disrupters are not very potent as compared to naturally occurring hormones. However, this is not always the case. Studies of the perinatal toxicity of TCDD indicate this toxicant is an extremely potent compound, producing effects in male offspring when administered on GD 15 at doses as low as 200 ng $kg^{-1}$ (Gray *et al.*, 1995b). In fact, our data indicate that TCDD is more potent in inducing malformations of the external genitalia in the female progeny than any synthetic or naturally occurring estrogen studied to date (especially estradiol). Other than TCDD, the only chemicals that have been shown to alter the development of external genitalia (clefting and hypospadias) of the female rat are potent estrogens like DES (Voherr *et al.*, 1979), RU 2858 (Vannier and Raynaud, 1980), estradiol (Vannier and Raynaud, 1980; Henry and Miller, 1986) and LY117018, which acts as an estrogen agonist in the fetal rat. At low doses, TCDD is more potent than any of these estrogens; the LOAEL for TCDD is well below the LOAEL for any of these estrogenic chemicals and the effects seen with low dosage levels of these estrogenic chemicals are comparable to the effects obtained with TCDD 1 μg $kg^{-1}$, administered on GD 15. For example, in female offspring exposed to TCDD on GD 15 the urethral opening is located near the caudal base of a cleft to partially cleft phallus but the urethral orifice is always separate from the vaginal orifice and the male offspring are not malformed. Similar effects were obtained when DES or estradiol were administered by injection on GD 19 (Henry *et al.*, 1984). DES 25 μg $kg^{-1}$ produced partial clefting of the phallus, without hypospadias (25-fold higher than our dose of TCDD) and while estradiol 25 000 μg $kg^{-1}$ was teratogenic, 2500 μg $kg^{-1}$ (2500-fold higher than our dose of TCDD) had no effect. With RU 2858, a dose about eight-fold higher than our dose of TCDD (2 μg per rat, administered at 0.4 μg per rat per d for 5 days, or about 8 μg $kg^{-1}$) produced a slight degree of hypospadias (no urethral–vaginal junction).

We have found that p,p′DDE is also a relatively potent antiandrogen, producing antiandrogenic effects in cells transfected with human androgen receptors at concentrations of 200 nm and above; effects nearly comparable to the active metabolite of the drug flutamide. In addition, one of the active metabolites of the antiandrogenic fungicide vinclozolin, M2, is of equivalent potency (Kelce *et al.*, 1994).

While it is often noted that the naturally occurring and man-made chemicals which exhibit estrogenic activity are less active than animal steroidal estrogens (Adams, 1989), this may not always be the case. For example, miroestrol, isolated from the woody climbing plant *Pueraria muerifica*, was as potent as 17β estradiol when administered subcutaneously to mice (Labov, 1977).

In addition to potency *in vitro* or *in vivo* from rodent studies, other equally important factors need to be considered in the risk assessment of endocrine disrupters, including exposure levels, bioavailability and metabolism. For example, in the assessment of the potential toxicity of an environmental estrogen to humans, one must consider whether or not the estrogenic toxicant binds to steroid hormone binding globulin (SHBG) proteins, which bind most of the estradiol in the blood, preventing uptake into target cells and tissues. Those that do not bind to SHBGs, like DES, may be more potent to the fetus than expected because they can readily enter the cells. In addition, although many environmental estrogens are 1000-fold less potent than estradiol (found at ppt in serum), they can occur at high concentrations in the diet and blood (ppm). Furthermore, although the concentration of a single chemical may not reach a toxic level, the effects of multiple chemical exposure to estrogens appear to be additive under certain conditions, and synergistic under others. For example, combined exposure to two estrogenic PCB metabolites altered sex differentiation in turtles to a much greater degree than expected (Bergeron *et al.*, 1994). In addition to external exposure levels, in the case of TCDD or DDE, these chemicals have long half-lives and bioaccumulate, such that continued exposure to low doses may eventually result in toxic tissue levels and in the case of lipophilic compounds like DDE and TCDD these are passed to the first-born child during gestation and lactation.

## 4.8 SCREENING FOR HORMONAL ACTIVITY

Currently, the only test protocols that expose animals during development and assess reproductive function of the offspring are the Multigenerational Reproductive Tests (reviewed by Palmer, 1981). At present, these protocols do not require either endocrine data or even reproductive organ weights, endpoints that reflect hormonal alterations. More recently, updated multigenerational test guidelines have been developed (Makris, 1995) which include indirect (bioassay) indices of endocrine function (pubertal landmarks, estrous cyclicity, reproductive organ weights, etc.), but these procedures have not yet been routinely implemented. In addition, our laboratory has developed an Alternate Reproduction Test (ART) protocol (Gray *et al.*, 1988b; Zenick *et al.*, 1994) which provides a comprehensive evaluation of the endocrine status of both the parental (P0) and subsequent (F1) generations. ART requires the measurement of serum hormone levels as well as reproductive organ weights. However, all of these long-term tests take more than a year to complete, they are expensive to conduct and shorter-term tests could be developed to identify endocrine disrupters more efficiently. A comprehensive screening strategy is needed because it is apparent that no single *in vivo* or *in vitro* test can adequately screen a compound for all of the mechanisms that are known to respond to endocrine disrupters. For example, screening chemicals *in vitro* only for estrogenicity would allow one to predict the *in vivo* alterations induced by DDE or vinclozolin.

We propose that an *in vivo*/*in vitro* test strategy be developed to screen chemicals for effects on the reproductive system. For example, we have found that dosing animals with toxicants during puberty is a simple means of detecting estrogenicity and antiandrogenicity. In these studies, male and female weanling rats are dosed with a chemical from 22 to 50 days of age and necropsied. The age and weight at puberty, the weights of the reproductive organs and serum hormones are measured. To date, this protocol has been used successfully with chlordecone, methoxychlor, octylphenol, vinclozolin and p,p′DDE. Pubertal alterations would also result from chemicals that disrupt hypothalamic-pituitary function, and, for this reason, additional *in vivo* and *in vitro* tests would be needed to identify the mechanism of action. Using the above protocol, if vaginal opening was accelerated then one could determine the ability of a chemical to acutely stimulate uterine growth. A compound that displayed uterotrophic activity could be examined *in vitro* for estrogen receptor binding; however, false negatives would result from this test if the compound required metabolic activation or was insoluble. Compounds that bound to the estrogen receptor would require future characterization *in vitro* to determine the degree to which they displayed agonist and/or antagonist activity and the nature of the interaction with HREs on the DNA.

The approach employed by Kelce *et al.* (1994) and Gray *et al.* (1994) is an example of how one would detect developmental reproductive toxicity of antiandrogenic compounds, like the active metabolites of vinclozolin. With vinclozolin, studies by the registrant suggested that this possessed antiandrogenic activity (van Ravenzwaay, 1992). In a developmental study (Gray *et al.*, 1994), vinclozolin produced antiandrogen-like alterations of male rat sex differentiation and, when administered for 30 days starting at weaning, it delayed puberty, reduced the weights of the sex accessory glands and increased serum testosterone and LH levels. Subsequently, it was determined *in vitro* that two metabolites of the parent compound bound competitively to the androgen receptor, but not the estrogen receptor (Kelce *et al.*, 1994). Additional *in vitro* studies determined that these metabolites bound to the human androgen receptor and displayed antiandrogenic activity.

In another case, where we suspected that DDE had antiandrogenic activity, based on effects seen in alligators (Guillette *et al.*, 1994), and because we knew this was not rapidly metabolized, DDE was initially tested *in vitro* for binding to the androgen receptor, rather than being initially tested *in vivo*. When positive results were obtained, we then conducted a series of *in vivo* and *in vitro* studies. We also propose that serum be taken and examined from these pubertal studies for alterations in the hormone levels, like the thyroid hormones, in order to determine if this component of the endocrine system is altered.

During testing, one must always keep in mind the limitations of each test and what the results truly mean. Problems with *in vitro* testing can result from lack of specificity of the test, cell viability, solubility, stability and metabolism, and for these reasons, one must confirm that the cells are alive and that the administered material is actually present in the medium at the expected concentrations. *In vivo* tests may also lack specificity and sensitivity. In addition, *in vitro* potency is often not a good predictor of potency *in vivo* for pharmacokinetic reasons. One additional screening method that is undergoing rapid development is the use of Quantitative Structure Activity Relationship (QSAR) models for the identification of toxicants that bind to steroid and Ah receptors (Waller and McKinney, 1995). Once QSAR models are developed, they can be used to screen libraries of chemical structures of potential hormonal activity. Although QSAR methods may eventually have great utility, they will always be susceptible to 'false negatives' because they will fail to predict hormonally active toxicants that have a structure that is 'outside' of the database used to develop the QSAR models.

## REFERENCES

Adams, R. R. (1989) Phytoestrogens, in *Toxicants of Plant Origin*. vol. IV: *Phenolics* (ed. P. R. Cheeke). CRC Press, Boca Raton, FL, pp. 23–51.

Allen, L. S., Hines, M., Shryne, J. E. and Gorski R. A. (1989) Two sexually dimorphic cell groups in the human brain. *J. Neurosci.*, **9**(2), 497–506.

Anderson, S. A., Fail, P. A., Pearce, S. W. *et al.* (1995) Testicular and adrenal response in adult male Long-Evans rats. *The Toxicologist*, **15**, 164.

Barter, A. and Klaassen, C. (1992) UDP-glucuronosyltransferase inducers reduce thyroid hormone levels in rats by an extrathyroidal mechanism. *Toxicol. Appl. Pharmacol.*, **113**, 36–92.

Bartomsky, C. (1974) Effects of polychlorinated biphenyl mixture (Aroclor 1254) and DDT on biliary thyroxine excretion on rats. *Endocrinology*, **95**, 1150–55.

Bartomsky, C., Murthy, P. and Banovac, K. (1976) Alterations in thyroxine metabolism produced by cutaneous application of microscope immersion oil: effects due to polychlorinated biphenyls. *Endocrinology*, **98**, 1309–14.

Battu, R. S., Singh, P. P., Joia, B. S. and Kalra, R. L. (1989) Contamination of bovine milk from indoor use of DDT and HCH in malaria control programmes. *Sci. Total Environ.*, **86**, 281–7.

Bengtsson, B. (1980) Long-term effects of PCB (Clophen A50) on growth, reproduction and swimming performance in the minnow (*Phoxinus phoxinus*). *Water Res.*, **14**, 681–7.

Bergeron, J. M., Crews, D. and McLachlan, J. A. (1994) PCBs as environmental estrogens: turtle sex determination as a biomarker of environmental contamination. *Environ. Health Perspect.*, **102**, 780–1.

Bern, H. A., Mills, K. T. and Jones, L. A. (1983) Critical period for neonatal estrogen exposure in occurrence of mammary gland abnormalities in adult mice. *Proc. Soc. Exp. Biol. Med.*, **172**, 239–42.

Birnbaum, L. S. (1994) Endocrine effects of prenatal exposure to PCBs, dioxins and other xenobiotics: implications for policy and research. *Environ. Health Perspect.*, **102**, 676–9.

Black, D. and McElroy A. (1991) Reproductive effects of 3,3′,4,4′-TCB exposure in *Fundulus heteroclitus*. 12th Annual SETAC, Seattle, WA.

Bouman, H., Reinecke, A. J., Cooppan, R. M. and Becker, P. J. (1990) Factors affecting levels of DDT and metabolites in human breast milk from Kwazulu. *J. Toxicol. Environ. Health*, **31**, 93–115.

Brezner, E., Terkel, J. and Perry, A. (1984) The effect of Aroclor 1254 (PCB) on the physiology of reproduction in the female rat-I. *Comp. Biochem. Physiol.*, **77**, 65.

Bryan, E. E., Gildersleeve, R. P. and Wiard, R. P. (1989) Exposure of Japanese quail embryos to o,p′-DDT has long-term effects on reproductive behaviors, hematology and feather morphology. *Teratology*, **39**, 525–35.

Burroughs, C. D., Bern, H. A. and Stokstad, E. L. (1985) Prolonged vaginal cornification and other changes in mice treated neonatally with coumestrol, a plant estrogen. *J. Toxicol. Environ. Health*, **15**, 51–61.

Carilla, E., Briley, M. Fauran, F. *et al.* (1984) Binding of Permixon, a new treatment for prostatic benign hyperplasia, to the cytosolic androgen receptor in the rat prostate. *J. Steroid Biochem.*, **20**, 521–3.

Carlsen, E., Giwercman, A., Keiding, N. and Skakkebaek, N. E. (1992) Evidence for decreasing quality of semen during past 50 years. *Br. Med. J.*, **305**, 609–13.

Casillas, E., Misitano, D., Johnson, L. *et al.* (1991) Inducibility of spawning and reproductive success of female English sole (*Paraphrsy vetulus*) from urban and nonurban areas of Puget Sound, WA. *Mar. Environ. Res.*, **31**, 99–122.

Chamness, G. C., Bannayan, G. A., Landry, L. A. Jr, *et al.* (1979) Abnormal reproductive development in rats after neonatally administered antiestrogen (Tamoxifen). *Biol. Reprod.*, **21**, 1087–90.

Clark, J. H. and McCormack, S. (1977) Clomid or nafoxidine administered to neonatal rats causes reproductive tract abnormalities. *Science*, **197**, 164–5.

Clevenger, M. A., Roberts, S. M., Lattin, D. L. *et al.* (1989) The pharmacokinetics of 2,2′,5,5′ tetrachlorobiphenyl and 3,3′,4,4′ tetrachlorobiphenyl and its relationship to toxicity. *Tox. Appl. Pharmacol.*, **100,** 315–27.

Colby, H. D. (1988) Adrenal gland toxicity: chemically induced dysfunction. *J. Am. Coll. Toxicol.*, **7,** 45–69.

Colborn, T. and Clement, C. (1992) *Chemically Induced Alterations in Sexual and Functional Development: the Wildlife/Human Connection (Advances in Modern Environmental Toxicology*, vol. 21), Princeton Scientific Publishing, Princeton, NJ.

Coussens, R. and Sierens, G. (1949) Estrogenic properties of tulip bulbs. *Arch. Int. Pharmacodyn. Ther.*, **78,** 309 (Chem. Abstr. 43 4727).

Davis, D. L., Bradlow, H. L., Wolff, M. *et al.* (1993) Medical hypothesis: xenoestrogens as preventable causes of breast cancer. *Environ. Health Perspect.*, **101,** 372–7.

Davis, W. P. and Bortone, S. A. (1992) Effects of mill effluent on fish sexuality, in *Advances in Modern Environmental Toxicology*, vol XXI: *Chemically Induced Alterations in Sexual and Functional Development: the Wildlife/Human Connection,* (eds T. Colborn and C. Clement). Princeton Scientific Publishing, Princeton, NJ, pp. 113–27.

Dewailly, E., Nantel, A., Weber, J. and Meyer, R. (1989) High levels of PCBs in breast milk of Inuit women from arctic Quebec. *Bull. Environ. Contam. Toxicol.*, **43,** 641.

Dohler, K. D., Coquelin, A., Davies, F. (1986) Pre- and postnatal influence of an estrogen antagonist on differentiation of the sexually dimorphic nucleus of the preoptic area of male and female rats. *Neuroendocrinology*, **42,** 443–8.

Ehrhardt, A. A. and Meyer-Bahlburg, F. L. (1981) Effects of prenatal sex hormones on gender-related behavior. *Science*, **211,** 1312–18.

Elliot, J. E., Martin, P. A., Arnold, T. W. and Sinclair, P. H. (1994) Organochlorines and reproductive success of birds in orchard and non-orchard areas of central British Columbia, Canada, 1990–1991. *Arch. Environ. Contam. Toxicol.*, **26,** 435–43.

Elskus, A. A. (1992) Polychlorinated biphenyl (PCB) effects, PCB congener distribution and cytochrome P450 regulation in fish. PhD dissertation, Boston University, Boston, Mass.

Emmett, E., Maroni, M. and Jeffreys N. (1988) Studies of transformer repair workers exposed to PCBs: II. Results of clinical laboratory investigations. *Am. J. Industr. Med.*, **14,** 47–62.

Ennis, B. W. and Davies, J. (1982) Reproductive tract abnormalities in rats treated neonatally with DES. *Am. J. Anat.*, **164,** 145–54.

Fail, P. A., Pearce, S. W., Anderson, S. A. *et al.* (1995) Endocrine and reproductive toxicity of vinclozolin (vin) in male Long-Evans hooded rats. *The Toxicologist*, **15,** 293.

Farnsworth, N. R., Bingel, A. S., Cordell, G. A. *et al.* (1975) Potential value of plants as sources of new antifertility agents II. *J. Pharm. Sci.*, **64,** 717.

Freeman, H. C. and Idler, D. R. (1975) The effect of polychlorinated biphenyl on steroidogenesis and reproduction in the brook trout (*Salvelinus fontinalus*). *Can. J. Biochem.*, **53,** 666–70.

Freeman, H. C., Sangalang, G. and Flemming, B. (1982) The sublethal effects of a polychlorinated biphenyl (Aroclor 1254) diet on the Atlantic cod (*Gadus morhua*). *Science Total Environ.*, **24,** 1–11.

Fry, D. M. and Toone, T. K. (1981) DDT-Induced feminization of gull embryos. *Science*, **213**(21), 922–4.

Gellert, R. J. (1978a) Kepone, Mirex, Dieldrin and Aldrin: estrogenic activity and the induction of persistent vaginal estrus and anovulation in rats following neonatal treatment. *Environ. Res.*, **16,** 131–8.

Gellert, R. J. (1978b) Uterotrophic activity of polychlorinated biphenyls and induction of precocious reproductive aging in neonatally treated female rats. *Environ. Res.*, **16,** 123–30.

Gellert, R. J. and Wilson, C. (1979) Reproductive function in rats exposed prenatally to pesticides and polychlorinated biphenyls (PCB). *Environ. Res.*, **18,** 437–43.

Gellert, R. J., Heinrichs, W. L. and Swerdloff, R. (1974) Effects of neonatally-administered DDT homologs on reproductive function in male and female rats. *Neuroendocrinology*, **16,** 84–94.

Giwercman, A., Carlsen, E., Keiding, N. and Skakkebaek, N. E. (1993) Evidence for increasing incidence of abnormalities of the human testis: a review. *Environ. Health Perspect.* (Suppl.), **101,** 65–71.

Golub, M., Donald, J. and Reyes, J. (1991) Reproductive toxicity of commercial PCB mixtures: LOAELS and NOAELS from animal studies. *Environ. Health Perspect.*, **94,** 245–53.

Gorski, R. A., Gordon, J. H., Shryne, J. E. and Southam, A. M. (1978) Evidence for a morphological sex difference within the medial preoptic area of the rat brain. *Brain Res.*, **148,** 333–46.

Goy, R. W. (1978) Development of play and mounting behavior in female rhesus virilized prenatally with esters of testosterone or dihydrotestosterone, in *Recent Advances in Primatology*, (eds D. J. Chivers, and J. Herbert), Academic Press, New York, pp. 449–62.

Gray, L. E. Jr (1982) Neonatal chlordecone exposure alters behavioral sex differentiation in female hamsters. *Neurotoxicology*, **3**(2), 67–80.

Gray, L. E. Jr (1992) Chemical-induced alterations of sexual differentiation: a review of effects in humans and rodents, in *Advances in Modern Environmental Toxicology*, vol. XXI: *Chemically Induced Alterations in Sexual and Functional Development: the Wildlife/Human Connection*, (eds T. Colborn and C. Clement), Princeton Scientific Publishing, Princeton, NJ, pp. 203–30.

Gray, L. E. Jr, and Ostby, J. S. (1995) *In utero* 2,3,7,8 tetrachlorodibenzo-p-dioxin (TCDD) alters reproductive morphology and function in female rat offspring. *Toxicol. Appl. Pharmacol.*, **131,** 285–94.

Gray, L. E. Jr, Ferrell, J. M. and Ostby, J. S. (1985) Alteration of behavioral sex differentiation by exposure to estrogenic compounds during a critical neonatal period: effects of zearalenone, methoxychlor and estradiol in hamsters. *Toxicol. Appl. Pharmacol.*, **80**(1), 127–36.

Gray, L. E. Jr, Ostby, J. S., Ferrell, J. M. *et al.* (1988a) Methoxychlor induces estrogen-like alterations of behavior and the reproductive tract in the female rat and hamster: effects on sex behavior, running wheel activity and uterine morphology. *Toxicol. Appl. Pharmacol.*, **96,** 525–40.

Gray, L. E. Jr, Ostby, J., Sigmonn, R. *et al.* (1988b) The development of a protocol to assess reproductive effects of toxicants in the rat. *Reprod. Toxicol.*, **2,** 281–7.

Gray, L. E. Jr, Ostby, J., Ferrell, J. *et al.* (1989) A dose-response analysis of methoxychlor-induced alterations of reproductive development and function in the rat. *Fund. Appl. Toxicol.*, **12,** 92–108.

Gray, L. E., Ostby, J. S., Linder, R. E. *et al.* (1990) Carbendazim-induced alterations of reproductive development and function in the rat and hamster. *Fund. Appl. Toxicol.*, **15,** 281–97.

Gray, L. E., Ostby, J., Marshall, R. and Andrews, J. (1993) Reproductive and thyroid effects of low-level polychlorinated biphenyl (aroclor 1254) exposure. *Fund. Appl. Toxicol.*, **20**, 288–94.

Gray, L. E., Ostby, J. S. and Kelce, W. R. (1994) Developmental effects of an environmental antiandrogen: the fungicide vinclozolin alters sex differentiation of the male rats. *Toxicol. Appl. Pharmacol.*, **129**, 46–52.

Gray, L. E., Kelce, W. R., Monosson, E. *et al.* (1995a) Exposure to TCDD during development permanently alters reproductive function in male LE rats and hamsters: reduced ejaculated and epididymal sperm numbers and sex accessory gland weights in offspring with normal androgenic status. *Toxicol. Appl. Pharmacol.*, **131**, 108–18.

Gray, L. E., Ostby, J. S., Wolf, C. *et al.* (1995b) Functional developmental toxicity of low doses of 2,3,7,8 tetrachlorodibenzo-p-dioxin and a dioxin-like PCB (169) in Long Evans rats and Syrian hamsters. Reproductive, behavioral and thermoregulatory alterations. In: *Dioxin* 95. 15th International Symposium on Chlorinated Dioxins and Related Compounds.

Guillette, L. J., Gross, T. S., Masson, G. R. *et al.* (1994) Developmental abnormalities of the gonad and abnormal sex hormone concentrations in juvenile alligators from contaminated and control lakes in Florida. *Environ. Health Perspect.*, **102**, 680–8.

Guo, Y. L., Lai, T. J., Ju, S. H. *et al.* (1993) Sexual developments and biological findings in Yucheng children, in *Dioxin '93. Organohalogen Compounds*, **14**, 235–8.

Guzelian, P. S. (1982) Comparative toxicology of chlordecone (Kepone) in humans and experimental animals. *Ann. Rev. Pharmacol. Toxicol.*, **22**, 89–113.

Hannon, W., Hill, F., Bernert, J. *et al.* (1987) Premature thelarche in Puerto Rico: a search for environmental estrogenic contamination. *Arch. Environ. Contam. Toxicol.*, **16**, 255–62.

Heinrichs, W. L., Gellert, R. J., Bakke, J. L. and Lawrence, N. L. (1971) DDT administered to neonatal rats induces persistent estrus syndrome. *Science*, **173**, 642–3.

Henry, E. C. and Miller, R. K. (1986) The antiestrogen LY117018 is estrogenic in the fetal rat. *Teratology*, **34**, 59–63.

Henry, E. C., Miller, R. K. and Baggs, R. B. (1984) Direct fetal injections of diethylstilbestrol and 17β-estradiol: a method for investigating their teratogenicity. *Teratology*, **19**, 297–304.

Hess, R. A., and Cooke, P. S. (1995) Neonatal thyroid status as one determinant of adult testis size. *Toxicologist*, **15**, 254.

Hines, M. and Green, R. (1991) Human hormonal and neural correlates of sex-typed behaviors. *Rev. Psychiatry*, **10**, 536–55.

Hines M. and Shipley, C. (1984) Prenatal exposure to diethylstilbestrol (DES) and the development of sexually dimorphic cognitive abilities and cerebral lateralization. *Develop. Psychol.*, **20**(1), 81–94.

Hong, C. S. and Bush, B. (1990) Determination of mono-and non-ortho coplaner PCBs in fish. *Chemosphere*, **21**, 173–81.

Huckins, J. N., Schwartz, T. R., Petty, J. D. and Smith, L. M. (1988) Determination, fate, and potential significance of PCBs in fish and sediment samples with emphasis on selected AHH-inducing congeners *Chemosphere*, **17**, 1995–2016.

Huseby, R. A. and Thurlow, S. (1982) Effects of prenatal exposure of mice to 'low-dose' diethylstilbestrol and the development of adenomyosis associated with the evidence of hyperprolactinemia. *Am. J. Obstet. Gynecol.*, **144**(8), 939–49.

Imperato-McGinley, J., Sanchez, R. S., Spencer, J. R. (1992) Comparison of the effects of the 5α-reductase inhibitor finasteride and the antiandrogen flutamide on prostate and genital differentiation: dose–response studies. *Endocrinology*, **131**, 1149–56.

Johnson, L. L., Casillas, E., Collier, T. K. *et al.* (1988) Contaminant effects on ovarian development in English sole (*Paraphrsy vetulus*) from Puget Sound, WA. *Can. J. Fish Aquat. Sci.*, **45,** 2133–45.

Jonsson, H., Keil, J., Gaddy, R. *et al.* (1976) Prolonged ingestion of commercial DDT and PCB: effects on progesterone levels and reproduction in the mature female rat. *Arch. Environ. Contam. Toxicol.*, **3,** 479.

Kalloo, N. B., Gearhart, J. P. and Barrack, E. R. (1993) Sexually dimorphic expression of estrogen receptors but not of androgen receptors in human fetal external genitalia. *J. Clin. Endocrinol. Metabol.*, **77,** 692–8.

Kelce, W. R., Monosson, E., Gamcsik, M. P. *et al.* (1994) Environmental hormone disruptors: evidence that vinclozolin developmental toxicity is mediated by antiandrogenic metabolites. *Toxicol. Appl. Pharmacol.*, **126,** 275–85.

Kelce, W. R., Stone, C. R., Laws, S. C. *et al.* (1995) The persistent DDT metabolite p,p′ DDE is a potent androgen receptor antagonist. *Nature*, **375,** 581–5.

Khera, K. S. and Ruddick, J. A. (1973) Polychlorodibenzo-p-dioxins: perinatal effects and the dominant lethal test in Wistar rats, in *Chlorodioxins–Origin and Fate* (ed. E. H. Blair). American Chemical Society, Washington, DC, pp. 70–84.

Koninckx, P. R., Braet, P., Kennedy, S. H. and Barlow, D. H. (1994) Dioxin pollution and endometriosis in Belgium. *Hum. Reprod.*, **9,** 1001–2.

Korach, K. S. (1994) Insights from the study of animals lacking functional estrogen receptors. *Science*, **266,** 1524–7.

Korach, K. S., Sarver, P., Chae, K., *et al.* (1987) Estrogen receptor-binding activity of polychlorinated hydroxybiphenyls: conformationally restricted structural probes. *Mol. Pharmacol.*, **33,** 120–6.

Kumagai, S. and Shimizu, S. (1982) Neonatal exposure to zearalenone causes persistent anovulatory estrus in the rat. *Arch. Toxicol.*, **50,** 270–86.

Labov, J. B. (1977) Phytoestrogens and mammalian reproduction. *Comp. Biochem. Physiol.*, **57,** 3–9.

Lai, T. J., Chou, Y. C., Chou, W. J. *et al.* (1993) Cognitive development in Yucheng children, in *Dioxin '93. Organohalogen Compounds*, **14,** 247–50.

Laws, S. C., Carey, S. A. and Kelce, W. R. (1995) Differential effects of environmental toxicants on steroid receptor binding. *The Toxicologist*, **15,** 294.

Leranth, C. S., Sequra, L. M. G., Palkovits, M. *et al.* (1985) The LH-RH containing neuronal network in the preoptic area of the rat: demonstration of LH–RH containing nerve terminals in synaptic contact with LH-RH neurons. *Brain Res.*, **345,** 332–6.

Le Vay, S. (1991) A difference in hypothalamic structure between heterosexual and homosexual men. *Science*, **253,** 1034–7.

Linder, R., Gaines, T. and Kimbrough, R. (1974) The effect of polychlorinated biphenyls on rat reproduction. *Food Cosmet. Toxicol.* **12,** 63.

Mably, T., Moore, R. W. and Peterson, R. E. (1992a) In utero and lactational exposure of male rats to 2,3,7,8-tetrachlorodibenzo-p-dioxin. 1. Effects on androgenic status. *Toxicol. Appl. Pharmacol.*, **114,** 97–107.

Mably, T., Moore, R. W., Goy, R. W. and Peterson, R. E. (1992b) In utero and lactational exposure of male rats to 2,3,7,8-tetrachlorodibenzo-p-dioxin. 2. Effects on sexual behavior and the regulation of luteinizing hormone secretion in adulthood. *Toxicol. Appl. Pharmacol.*, **114,** 108–17.

Mably, T., Bjerke, D. L., Moore, R. W. *et al.* (1992c) In utero and lactational exposure of male rats to 2,3,7,8-tetrachlorodibenzo-p-dioxin. 3. Effects on spermatogenesis and reproductive capability. *Toxicol. Appl. Pharmacol.*, **114,** 118–26.

Makris, S. (1995) Proposed revisions to the testing guidelines for developmental toxicity and two-generation reproductive toxicity studies conducted under FIFRA and TSCA-A progress update. *The Toxicologist*, **15**, 287.

McEwen, B. S. (1980) Gonadal steroid and brain development. *Biol. Reprod.*, **22**, 43–8.

McGlone, J. (1980) Sex differences in human brain asymmetry: a critical survey. *Behavior Brain Sci.*, **3**, 215–63.

McKinney, J. (1987) Do residue levels of polychlorinated biphenyls (PCBs) in human blood produce mild hypothyroidism? *J. Theor. Biol.*, **129**, 231.

McLachlan, J. A. (1981) Rodent models for perinatal exposure to diethylstilbestrol and their relation to human disease in the male, in *Developmental Effects of Diethylstilbestrol in Pregnancy* (eds A. L. Herbst and H. A. Bern), New York, Thieme-Stratton, pp. 48–157.

McLachlan, J. A., Newbold, R. R., Shah, H. C. *et al.* (1982) Reduced fertility in female mice exposed transplacentally to diethylstilbestrol (DES). *Fertility and Sterility*, **38**(3), 364–71.

Meyer-Bahlburg, H. F. L., Ehrhardt, A. A., Feldman, J. F. (1985) Sexual activity level and sexual functioning in women prenatally exposed to diethylstilbestrol. *Psychosom. Med.*, **47**(6), 497–511.

Monosson, E. and Stegeman, J. J. (1994) Induced cytochrome P4501A in winter flounder *Pleuronectes americanus*, from offshore and coastal sites. *Can. J. Fish. Aquat. Sci.*, **51**, 933–41.

Monosson, E., Fleming, W. J. and Sullivan, C. V. (1994) Effects of the planar PCB 3,3′,4,4′-tetrachlorobiphenyl (TCB) on ovarian development, plasma levels of sex steroid hormones and vitellogenin, and progeny survival in the white perch (*Morone americana*). *Aq. Tox.* **29**, 1–19.

Monosson, E., Kelce, W. R., Gray, L. E. (in preparation) Effects of perinatal exposure to the fungicide vinclozolin on prepubertal male rats.

Murray, F. J., Smith, F. A., Nitschke, K. D. *et al.* (1979) Three-generation reproduction study in rats given 2,3,7,8-tetrachlorodibenzo-p-dioxin (TCDD) in the diet. *Toxicol. Appl. Pharmacol.*, **50**, 241–52.

National Research Council (1993) *Pesticides in the Diets of Infants and Children.* National Research Council. National Academy Press, Washington, DC.

Newbold, R. R. (1993) Gender-related behavior in women exposed prenatally to diethylstilbestrol. *Environ. Health Perspect.*, **101**, 208–13.

Newbold, R. R. and McLachlan, J. J. (1985) Diethylstilbestrol-associated defects in murine genital tract development, in *Estrogens in the Environment II: Influences on Development* (ed. J. J. McLachlan), Elsevier North-Holland, New York, pp. 288–318.

Newbold, R. R., Bullock, B. C. and McLachlan, J. A. (1987) Mullerian remnants of male mice exposed prenatally to diethylstilbestrol. *Teratog. Carcinogen. Mutagen*, **7**, 377–89.

Niimi, A. J. and Oliver, B. G. (1983) Biological half-lives of polychlorinated biphenyl (PCB) congeners in whole fish and muscle of rainbow trout (*Salmo gairdneri*). *Can. J. Fish. Aquat. Sci.*, **40**, 1388–94.

NIOSH (1977) Criteria for a Recommended Standard. Occupational Exposure to Polychlorinated Biphenyls (PCBs). Rockville, MD, US DHEW, PHS, CDC. NIOSH Publ. No. 77–225.

Orberg, J., and Ingvast, C. (1976) Effects of pure chlorobiphenyls (2,4′,5-trichlorobiphenyl or 2,3′,4,4′,5,5′-hexachlorobiphenyl) on the disappearance of C14 from the blood plasma after intravenous injection of 4-C14 -progesterone and

on the hepatic drug metabolizing system in the female rat. *Acta Pharmacol. Toxicol.*, **41,** 11.
Palmer, T. (1981) Regulatory requirements for reproductive toxicology: theory and practice, in *Developmental Toxicology* (eds. C. Kimmel and J. Buelke-Sam), Raven Press, NY, pp. 259–88.
Peakall, D. (1970) p,p′DDT: effect on calcium metabolism and concentration of estradiol in the blood. *Science*, **168,** 592–4.
Poland, A., Smith, D., Kuntzman, R., *et al.* (1970) Effect of intensive occupational exposure to DDT on phenylbutazone and cortisol metabolism in human subjects. *Clin. Pharmacol. Ther.*, **11,** 724–32.
Polani, P. E. (1981) Abnormal sex development in man. 1. Anomalies of sex determining mechanisms, in *Mechanisms of Sex Differentiation in Animals and Man* (eds R. Austin and R. G. Edwards), Academic Press, New York, pp. 465–547.
Pomerantz, S. M., Roy, M. M., Thornton, J. E. and Goy, R. W. (1985) Expression of adult female patterns of sexual behavior by male, female and pseudohermaphroditic female rhesus monkeys. *Biol. Reprod.*, **33,** 878–89.
Purdom, C. E., Hardiman, P. A., Bye, V. J. (1994) Estrogenic effects of effluents from sewage treatment works. *Chem. Ecol.*, **8,** 275–85.
Sager, D. (1983) Effect of postnatal exposure to polychlorinated biphenyls on adult male reproductive function. *Environ. Res.*, **31,** 76.
Sager, D. B. and Girard, D. M. (1994) Long-term effects on reproductive parameters in female rats after translactational exposure to PCBs. *Environ. Res.*, **66,** 52–76.
Sager, D. B., Shih-Schroeder, W. and Girard, D. (1987) Effect of early postnatal exposure to polychlorinated biphenyls (PCBs) on fertility in male rats. *Bull. Environ. Contam. Toxicol.*, **38,** 946–53.
Sager, D., Girard, D. and Nelson, D. (1991) Early postnatal exposure to PCBs. Sperm function in rats. *Environ. Toxicol. Chem.*, **10,** 737–46.
Salunkhe, D. K., Adsule, R. N. and Bhonsle, K. I. (1989) Antifertility agents of plant origin, in *Toxicants of Plant Origin.* Volume IV: *Phenolics* (ed. P. R. Cheeke), CRC Press, Boca Raton, FL, pp. 53–81.
Schardein, J. L. (1993) Hormones and hormonal antagonists, in *Chemically Induced Birth Defects*, Marcel Dekker, New York, pp. 271–339.
Sharp, J. R. and Thomas, P. (1991) Influence of maternal PCB exposure on embryonic development of the Atlantic croaker *Micropogonias undulatus.* 12th Annual SETAC, Seattle, WA.
Sivarajah, K., Franklin, C. S. and Williams, W. P. (1978) The effects of polychlorinated biphenyls on plasma steroid levels and hepatic microsomal enzymes in fish. *J. Fish. Biol.*, **13,** 401–9.
Smits-van Prooije, A. E., Lammers, J, Waalkens-Berendsen, D. H. *et al.* (1993) Effects of the PCB 3,4,5,3′,4′,5′ hexachlorobiphenyl on the reproduction capacity of Wistar rats. *Chemosphere*, **27,** 395–400.
Spies, R., and D. W. Rice, Jr., 1988. The effect of organic contaminants on reproduction in the sarry flounder, *Platichthys stellatus* (Pallas) in San Francisco Bay. II. Reproductive success of fish captured in San Francisco Bay and spawned in the laboratory. *Mars. Biol. Res.* **98,** 191–200.
Spitzer, P. R., Risebrough, R., Walker, W. *et al.* (1978) Productivity of ospreys in Connecticut–Long Island increases as DDE residues decline. *Science*, **202,** 333–5.
Stanczyk, F. Z. (1994) Structure-function relationships, potency and pharmacokinetics of progestogens, in *Treatment of Postmenopausal Women: Basic and Clinical Aspects* (ed. R. A. Lobo), Raven Press, New York, pp. 69–89.

Steinberger, E. and Lloyd, J. A. (1985) Chemicals affecting the development of reproductive capacity, in *Reproductive Toxicology* (ed. R. L. Dixon), New York, Raven Press, pp. 1–20.

Swaab, D. F. and Fliers, E. A. (1985) Sexually dimorphic nucleus in the human brain. *Science*, **228,** 1112–14.

Tanabe, S., Kannen, N., Subramanian, A. *et al.* (1987) Highly toxic coplanar PCBs: occurrence, source, persistency and toxic implications to wildlife and humans. *Environ. Pollution*, **47,** 147–63.

Tarhanan, J., Koistinenen, J., Paasivirta, J. *et al.* (1989) Toxic significance of planar aromatic compounds in Baltic ecosystem – new studies on extremely toxic coplanar PCBs. *Chemosphere*, **18,** 1067–77.

Thomas, P. (1988) Reproductive endocrine function in female Atlantic croaker exposed to pollutants. *Mar. Environ. Res.*, **24,** 179–83.

Tilson, H., Jacobson, J. and Rogan, W. (1990) Polychlorinated biphenyls and the developing nervous system: cross-species comparisons. *Neurotoxicol. Teratol.*, **12,** 239.

van Ravenzwaay, B. (1992) Discussion of prenatal and reproductive toxicity of Reg. No. 83–258 (Vinclozolin). Data submission to USEPA from BASF Co., MRID 425813–02.

Vannier, B. and Raynaud, J. P. (1980) Long-term effects of prenatal oestrogen treatment on genital morphology and reproductive function in the rat. *J. Reprod. Fertil.*, **59,** 43–9.

Vodicknik, M. J., and Peterson, R. E. (1985) The enhancing effect of spawning on elimination of a persistent polychlorinated biphenyl from female yellow perch. *Fund. Appl. Toxicol.*, **5,** 770–6.

von Westernhagen, H., Rosenthal, H., Dethlefsen, V. *et al.* (1981) Bioaccumulating substances and reproductive success in Baltic flounder *Platichthys flesus*. *Aquat. Toxicol.*, **1,** 85–99.

Voherr, H., Messer, R. H., Voherr, U. F. *et al.* (1979) Teratogenesis and carcinogenesis in rat offspring after transplacental and transmammary exposure to diethylstilbestrol. *Biochem. Pharmacol.*, **28,** 1865–77.

Walker, M. K. and Peterson, R. E. (1992) Toxicity of polychlorinated dibenzo-p-dioxins, dibenzofurans and biphenyls during early development in fish, in *Advances in Modern Environmental Toxicology*, vol. XXI: *Chemically Induced Alterations in Sexual and Functional Development: the Wildlife/Human Connection* (eds T. Colborn and C. Clement), Princeton Scientific Publishing, Princeton, NJ, pp. 195–203.

Waller, C. and McKinney, J. (1995) Development, validation and application of three-dimensional quantitative structure activity relationship model for dioxins and related polycyclic compounds. *The Toxicologist*, **15,** 271.

Wannemacher, R., Rebstock, A., Kulzer, E., *et al.* (1992) Effects of 2,3,7,8-tetrachlorodibenzo-*p*-dioxin on reproduction and oogenesis in zebrafish (*Brachydanio rario*). *Chemosphere*, **24,** 1361–8.

Whitsett, J. M. and Vandenbergh, J. G. (1975) Influence of testosterone propionate administered neonatally on puberty and bisexual behavior in female hamsters. *J. Comp. Physiol. Psychol.*, **88,** 248–55.

Wilson, J. D. (1978) Sexual differentiation. *Ann. Rev. Physiol.*, **40,** 279–306: **16,** 123–30.

Wilson, J. D. (1992) Syndromes of androgen resistance. *Biol. Reprod.*, **46,** 168–73.

Zenick, H., Clegg, E. D., Perreault, S. D. *et al.* (1994) Assessment of male reproductive toxicity: a risk assessment approach, in *Principle and Methods of Toxicology*, 3rd edn (ed. A. W. Hayes), Raven Press, New York, pp. 937–88.

# PART TWO

## *Empirical Evidence for Linkages*

# 5 *The Great Lakes: a model for global concern*

THEO COLBORN

Questions concerning the long-term, delayed effects of transgenerational exposure to xenobiotics led to the gathering of a diverse group of international scientists to discuss 'Chemically Induced Alterations in Sexual Development: the Wildlife/Human Connection'. The experts, representing 17 disciplines from anthropology to zoology, met at the Wingspread Conference Center, on the shore of Lake Michigan, Racine, Wisconsin, in July 1991. Sharing their experiences in basic science and applied science they discussed their findings from years of work in wildlife toxicology and behavior to human epidemiology in pharmacology and psychology. For many in attendance this eclectic approach was a new experience, having never before attended a scientific meeting outside their discipline. It was also a new experience in that the participants were asked to come to some agreement in principle concerning the magnitude and scope of the problem of xenobiotic-induced alterations in sexual and functional development (Colborn and Clement, 1992).

Increased interest in the implications of developmental effects on future generations of wildlife and humans dates back to an extensive literature review on the status of wildlife and human health in the Great Lakes basin (Colborn *et al.*, 1990). Populations of birds, fish, reptiles and mammals in the region had exhibited declines or were suffering instability because of a suite of effects expressed as damage to the endocrine, nervous, and/or immune systems primarily in the offspring. Adult animals appeared relatively healthy reflecting recruitment from outside the Great Lakes region as adult recruitment was poor or absent from within the region. However, linking these problems to contaminants was difficult because of the functional nature of the effects and the delay between parental exposure and expression of the effects in the offspring. What this meant in terms of human health had been

*Interconnections Between Human and Ecosystem Health.* Edited by Richard T. Di Giulio and Emily Monosson. Published in 1996 by Chapman & Hall, London. ISBN 0 412 62400 1.

troubling public health authorities, policymakers and citizens for a number of years (National Research Council of the United States and Royal Society of Canada, 1985). This concern contributed to the need for Wingspread.

Before the meeting closed the Wingspread participants concluded that widespread human exposure to various endocrine-disrupting pesticides and industrial chemicals is not merely possible, it is probable, and that 'unless the environmental load of synthetic hormone disruptors is abated and controlled, large-scale dysfunction at the population level is possible' (Colborn and Clement, 1992).

They called for integrated cooperative research to develop both wildlife and laboratory models for extrapolating risks to humans. They also noted that 'The effects of endocrine disruptors on longer-lived humans may not be as easily discerned as in shorter-lived laboratory or wildlife species', and therefore called for 'early detection methods – to determine if human reproductive capability is declining'.

These statements were prompted by the parallels that became evident during the meeting as the participants described their research – parallels between the adverse developmental effects reported in wild animals, laboratory animals, humans and *in vitro* studies where exposure involved the same chemicals or chemicals with similar mechanisms of action. In each case, the effects were associated with exposure to chemicals capable of disrupting the endocrine system (Colborn and Clement, 1992). For example, wildlife specialists reported on endocrine abnormalities in wild animals that were first recognized in the early 1950s. In some cases, notably birds (Fry and Toone, 1981) and common seals (Reijnders, 1986), these anomalies were later replicated in confined wild animals exposed to the same contaminants at concentrations found in the affected wild birds and seals. Next, basic scientists described the exquisite sensitivity of the endocrine system to slight shifts in natural female/male hormone ratios (vom Saal, 1983) and toxicologists revealed the delayed, generally invisible, long-term effects of very low doses of synthetic chemicals on the endocrine system in laboratory animals (Peterson *et al.*, 1992), effects similar to those described by the basic scientists when hormone ratios of endogenous hormones were skewed. And last, human reproductive physiologists described the anomalies induced in laboratory animals and humans exposed in the uterus to the synthetic hormone diethylstilbestrol (DES).

The parallels were remarkable – within systems and among species. At each level of biological organization, from the molecular to the cellular, to tissue, organ, whole animal and even at the population level, similar conditions were reported in wild animals, laboratory animals and humans exposed to chemicals known to disrupt the endocrine system. The similarities across species and at each level of biological organization were of greater concern to those in attendance at Wingspread than the lack of definitive cause and effect linkages in wildlife and human populations.

Participants also acknowledged that following exposure to endocrine-disrupting chemicals the effects were seldom expressed immediately in the adult, but were evident in the offspring, and often not until the offspring reached adulthood. This was borne out by evidence that annually large numbers of migratory birds are seen nesting around the Great Lakes. However, their reproductive success is poor and adult recruitment comes more from immigration rather than from within the nesting population (Colborn, 1991b). Whether in the wild or the laboratory, the fetus was extremely vulnerable to exogenous chemicals, which by postnatal criteria were weakly active. This was evident in humans by the lack of measurable effects in the mothers who took DES, but whose children have histories of developmental problems. The long-term changes in human and mouse reproductive tracts that resulted from exposure to the endocrine disruptor DES could also serve as a model for other man-made endocrine-disrupting chemicals and highlighted the sensitivity of the developing endocrine, immune, and nervous systems to exposure of this nature (Bern, 1992). This was affirmed recently when the female pups of pregnant rats, fed a single meal of dioxin (1 μg kg $bw^{-1}$) on day 15 of gestation, exhibited similar cellular, tissue and morphological changes of the reproductive tract as rats, mice and hamsters exposed to DES *in utero* (Gray and Ostby, 1995).

Many of the endocrine problems described at the Wingspread meeting were first recognized by wildlife biologists in the Great Lakes basin among top predator species of birds, fishes, mammals and reptiles (Colborn *et al.*, 1990). Until recently, it was thought that these problems were unique to the region. However, it is now realized that the Lakes, holding 20% of the world's fresh water, are a model for the world.

Following the Second World War the Great Lakes region became home for one of the largest industrial and agro-chemical complexes in North America (Colborn *et al.*, 1990). The Lakes and surrounding lands also became major disposal sites for the wastes produced by these industries and other human activities. For a number of years it was assumed that the bountiful water resources of the region were capable of assimilating these chemicals and maintaining the integrity of the system. However, reports started appearing in the scientific literature in the early 1970s indicating that all was not well within the Great Lakes wildlife community (Gilbertson and Hale, 1974; Ludwig, 1974). Even today, after years of attempting to regulate the release of contaminants into the Great Lakes environment, 16 top predator species still exhibit reproductive problems and are unable to maintain stable populations (Leatherland, 1992; Bowerman *et al.*, 1995). In some locations entire populations have been extirpated. Early mortality among embryos (in the egg or the womb) and loss of fertility among surviving offspring contribute to the population problems (Gilbertson and Schneider, 1991).

The anomalies reported in the wildlife offspring that lead to premature death or infertility have been associated with the offspring's indirect exposure, through the mother, to a mixture of chemicals that biomagnify in the Great Lakes food web. Chemicals that were once considered safe because they were not acute toxicants, powerful carcinogens, or mutagens at ambient concentrations had penetrated the Great Lakes ecosystem in large volumes and became widely dispersed. Only over the past decade was it discovered that some of these 'safe' compounds were capable of mimicking or interfering with naturally produced hormones, neurotransmitters, growth factors and inhibiting substances (Colborn *et al.*, 1993). In this role as agonists, antagonists, modulators and blockers, they change the course of development of the embryo, fetus and newborn by interfering with normal gene expression and differentiation. The resulting loss of function, such as loss of fertility, is not easy to recognize nor is it easy to link with a specific compound because so many chemicals are now present in the environment.

Although high-volume, direct dumping into the Lakes has been significantly curtailed since the early 1970s, other activities that contribute to the contamination of the Lakes are more intractable. Non-point source contamination from heavy agri-business activity and long- and short-range air transport contributions have not been resolved (US EPA, 1994). For example, both runoff and atmospheric deposition have been shown to contribute to the herbicide atrazine in the Great Lakes. Atrazine has been reported at 115 ppt from the surface to the bottom of the water column in Lake Erie (Eisenreich and Schottler, 1994). Mass-balance studies reveal that long-range atmospheric transport contributes 90% of the PCB loading to Lake Superior. PCBs are delivered by air to Arctic Canada at the same rate they are delivered to Lake Superior. Most important, the insecticide lindane is deposited in the Arctic at 32.2 ng/m$^2$, almost 100 times greater than atmospheric deposition of lindane to the Great Lakes (Barrie *et al.*, 1992). The Great Lakes are not the only recipients of airborne contaminants.

Endocrine-disrupting chemicals in the environment are a global problem, not limited to one geographic area alone (Colborn, 1991a; Muir *et al.*, 1992). Because of the early research in the Great Lakes region of North America, however, the Lakes have been used as models for the impact of xenobiotics on wildlife and human health. A second Wingspread Work Session, convened in December 1993 to address the 'Environmentally Induced Alterations in Development: a Focus on Wildlife', attended this time by wildlife biologists from the North American continent, concurred that:

> Declines in a number of species and taxa (including plants) are in progress on the North American continent. Some of these declines are related to exposure to man-made chemicals. Such declines are not solely a US or North American problem but are occurring on a global scale. . . . Populations of many long-lived species are declining, some to the verge of extinction, without society's knowledge. The presence of breeding adults and even healthy young does not necessarily reflect a healthy population.

Detailed population analysis is needed to determine whether offspring have the functional capacity to survive reproduction.

They estimated with confidence that: 'In many cases wildlife and humans have exceeded their capacity to compensate for exposure to chemicals' (Consensus Statement, 1995).

Worldwide, future generations will be faced with the problems created by (1) the pervasive contamination of the environment by persistent compounds already released that now contribute to a background body burden that has reached relevant levels for concern, (2) the increasing production and use of new agricultural chemicals that may be shorter-lived but require only one 'hit' during embryonic development to have a lasting impact, and (3) the continuing use and introduction of new industrial chemicals into commerce, many of which were once considered benign, but are now known to be biologically active.

## REFERENCES

Barrie, L., Gregor, D., Hargrave, B. *et al.* (1992) Arctic contaminants: sources, occurrence and pathways. *Science and the Total Environment*, **122,** 1–74.

Bern, H. (1992) The fragile fetus, in *Chemically Induced Alterations in Sexual and Functional Development: the Wildlife/Human Connection* (eds T. Colborn and C. Clement), Princeton Scientific Publishing, Princeton, NJ.

Bowerman, W., Giesy, J., Best, D. *et al.* (1995) A review of factors affecting productivity of bald eagles in the Great Lakes region: implications for recovery. *Environmental Health Perspectives* (Suppl.), **103**(4), 51–9.

Colborn, T. (1991a) Global implications of Great Lakes wildlife research. *International Environmental Affairs,* **3**(1), 3–25.

Colborn, T. (1991b) Epidemiology of Great Lakes bald eagles. *Journal of Toxicology and Environmental Health*, **33**(4), 395–454.

Colborn, T. and Clement, C. (eds) (1992) *Chemically Induced Alterations in Sexual and Functional Development: the Wildlife/Human Connection* (Advances in Modern Environmental Toxicology, vol. 21), Princeton Scientific Publishing, Princeton, NJ.

Colborn, T., Davidson, A., Green, S. *et al.* (1990) *Great Lakes, Great Legacy?* The Conservation Foundation, Washington, DC.

Colborn, T., vom Saal, F. and Soto, A. (1993) Developmental effects of endocrine-disrupting chemicals in wildlife and humans. *Environmental Health Perspectives*, **101**(5), 378–84.

Consensus Statement (1995) Environmentally induced alterations in development: a focus on wildlife. *Environmental Health Perspectives* (Suppl.), **103**(4), 3–5.

Eisenreich, S. and Schottler, S. (1994) Herbicides in the Great Lakes. *Environmental Science Technology*, **28,** 2228–32.

Fry, D. M. and Toone, C. K. (1981) DDT-induced feminization of gull embryos. *Science*, **213,** 922–4.

Gray, L. E. and Ostby, J. (1995) *In Utero* 2, 3, 7, 8-tetrachlorodibenzo-p-dioxin (TCDD) alters reproductive morphology and function in female rat offspring. *Toxicology and Applied Pharmacology*, **133,** 285–94.

Gilbertson, M. and Hale, R. (1974) Early embryonic mortality in a colony of herring gulls in Lake Ontario. *Canadian Field Naturalist*, **88,** 354–6.

Gilbertson, M. and Schneider, S. (1991) Special Issue: International Joint Commission Workshop on Cause–Effect Linkages. *Journal of Toxicology and Environmental Health*, **33**(4), 359–640.

Leatherland, J. (1992) Endocrine and reproductive function, in *Chemically Induced Alterations in Sexual and Functional Development: the Wildlife/Human Connection* (eds. T. Colborn and C. Clement), Princeton Scientific Publishing, Princeton, NJ.

Ludwig, J. (1974) Recent changes in the ring-billed gull population and biology in the Laurentian Great Lakes. *Auk*, **91,** 575–94.

Muir, D., Wagemann, R., Hargrave, B. *et al.* (1992) Arctic marine ecosystem contamination. *The Science of the Total Environment*, **122,** 75–134.

National Research Council of the United States and the Royal Society of Canada (1985) *The Great Lakes Water Quality Agreement. An Evolving Instrument for Ecosystem Management*. National Academy Press of the National Academy of Sciences.

Peterson, R., Moore, R., Mably, T. *et al.* (1992) Male reproductive system ontogeny: effects of perinatal exposure to 2,3,7,8-tetrachlorodibenzo-p-dioxin, in *Chemically Induced Alterations in Sexual and Functional Development: the Wildlife/Human Connection* (eds T. Colborn and C. Clement), Princeton Scientific Publishing, Princeton, NJ, pp. 175–93.

Reijnders, P. J. H. (1986) Reproductive failure in common seals feeding on fish from polluted coastal waters. *Nature*, **324,** 456–7.

vom Saal, F. (1983) The interaction of circulating estrogens and androgens in regulating mammalian sexual differentiation, in *Hormones and Behavior in Higher Vertebrates* (eds J. Balthazart, E. Prove and R. Giles), Springer-Verlag, Berlin, pp. 159–77.

US EPA (1994) Deposition of Air Pollutants to the Great Waters: First Report to Congress. Office of Air Quality Planning and Standards. EPA-453/R-93-055.

6

# Establishing possible links between aquatic ecosystem health and human health: an integrated approach*

S. MARSHALL ADAMS and MARK S. GREELEY, JR

## 6.1 INTRODUCTION

The effects of environmental contaminants on biological systems are well documented, particularly in aquatic ecosystems (Adams, 1990; McCarthy and Shugart, 1990; Boudou and Ribeyre, 1993). In addition, the effects of toxicants on human health, including those chemicals that are mediated through ingestion and both water- and air-borne pathways, have also been extensively studied (Eisenreich *et al.*, 1981; Rapaport *et al.*, 1985; Safe, 1987; Kannan *et al.*, 1988; Anderson, 1989). Potential relationships and links, however, between aquatic system health and human health are poorly understood because of problems related to:

1. establishing cause and effect between receptor systems and contaminants in sentinel ecosystem species and humans;
2. establishing common mechanisms of toxic action between sentinel species and humans;
3. identifying routes or pathways of exposure from the affected natural system (aquatic system) to human populations;
4. absence, in many cases, of appropriate epidemiological or similar studies on human populations.

Given that appropriate epidemiological data are not available in most cases, alternative approaches are needed that could help identify possible causal

*Publication No. 4360, Environmental Sciences Division, Oak Ridge National Laboratory.

*Interconnections Between Human and Ecosystem Health*. Edited by Richard T. Di Giulio and Emily Monosson. Published in 1996 by Chapman & Hall, London. ISBN 0 412 62400 1.

relationships between ecosystem and human health. The purpose of this study, therefore, is to develop and apply an integrated approach for establishing possible relationships between the health of aquatic systems and human health. To accomplish this objective, we will use as a test case example a river contaminated by bleached kraft mill effluents (the major biological contaminants of concern being dioxin and chlorophenolic compounds: EPA, 1988; Blevins, 1991) where the fishery resources in this system are known to be severely impacted by the pulp mill discharges (Adams *et al.*, 1992).

## 6.2 APPROACH

To develop and apply such a model approach for establishing possible links and relationships between ecosystem and human health, a four-step process will be taken (Figure 6.1). These steps are:

1. measure and document, at several levels of biological organization, the effects of contaminants on indicator fish species in the aquatic system being studied;
2. establish potential links in terms of how contaminants can be and are transferred from the aquatic system to the human population;
3. identify the common modes of action or receptors in both the sentinel species in the affected ecosystem and in the human population;
4. suggest and/or provide evidence for possible implications to human health as a result of exposure to these contaminants.

The first of the four-step process, establishing effects on the important fishery resources due to contaminant exposure, has been demonstrated, in part, from the studies of Adams *et al.* (1992). This study reported that a variety of effects occurred on the fish population and communities as a result of exposure to pulp mill effluents. The second component of this study will involve establishing potential links in terms of how contaminants are transferred from the aquatic system in question to the human population. Potential pathways which apply to this case are water contact, ingestion (drinking water and fish consumption), irrigation of crops and livestock watering.

Identifying common modes of action and receptors in the sentinel species and in humans is the third step involved in developing this integrated approach. Much of the information needed to address this aspect of the study, however, is available from the literature. The principal target areas where the actions of toxicants such as chlorinated hydrocarbons produce the greatest effects and concerns relative to human health (given of course that there are links or transfers of contaminants into the human population as established in step 2 above) are endocrine and reproductive system dysfunction, carcinogenic effects, immune system impairment and behavioral disorders (Colborn and Clement, 1992; Colborn *et al.*, 1993).

In the final phase of this model development process, implications to human

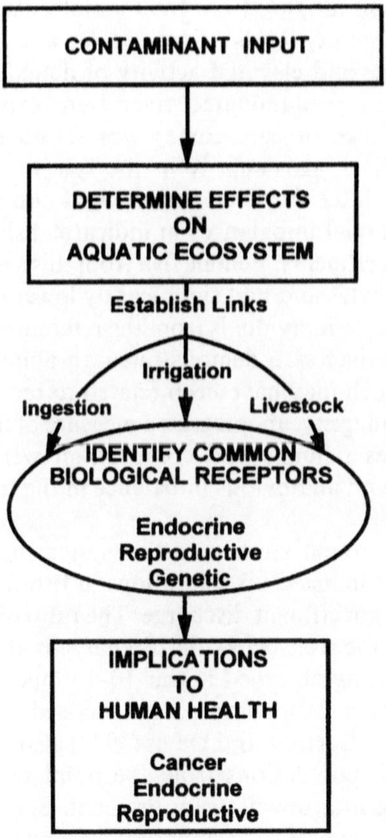


**Figure 6.1** Integrated four-step process for establishing possible links and relationships between ecosystem and human health.

health will be addressed in terms of providing evidence that suggest possible cause and effect relationships between environmental and human health. In the absence of definitive epidemiological studies, which are not feasible in many cases, we will attempt to provide alternative lines of evidence, such as the weight of evidence strategy used in ecological risk assessment (Suter, 1993), which could suggest causal relationships between contaminant effects on aquatic resources and human health effects.

## 6.3 DEVELOPMENT OF MODEL COMPONENTS

### 6.3.1 *DOCUMENTATION OF ECOSYSTEM HEALTH EFFECTS*

Studies by Adams *et al.* (1992) on a river system impacted by bleached kraft mill effluents demonstrated effects on the fishery resources at several levels of biological organization ranging from the biomolecular (genetic) level to the population and community levels. Biomolecular and biochemical responses such as DNA damage and elevated activity of detoxification enzymes indicated that fish in the contaminated river were exposed to toxicants. A relatively high incidence of carcinomas was found in the organs of fish sampled from this river, particularly in the spleen. The status of various condition indices such as the liver and visceral somatic indices suggested metabolic and nutritional imbalances in indicator fish species as a result of exposure to pulp mill effluents. Female fish from this system contained a large number of atretic oocytes and had significantly lower serum levels of steroid hormones than did those individuals from the reference site. Fish populations in the contaminated river also demonstrated an abnormal size distribution and age structure which may have been related to reproductive impairment. The Index of Biotic Integrity, an integrated measure of fish community health, indicated that species richness and composition were much lower in the contaminated river with an obvious imbalance in the trophic structure of the fish community.

In addition, some recent studies on this system have found evidence of possible contaminant-induced sex alterations in fish for several miles downstream from the point of effluent discharge. The ratio of male to female fish in the upper sections of the river below the discharge is at least 99 to 1 with only 1 or 2 females ever being observed in this 10–15 mile reach. The midsection and lower areas of the river, however, are composed of fish populations with near normal sex ratios. Bortone and Davis (1994) also observed an altered sex ratio in mosquitofish populations from the point of a paper mill effluent discharge proceeding in a downstream gradient. Sex alterations have been reported in organisms exposed to various types of contaminants including pulp mill effluents (Caruso *et al.*, 1988; Gibbs *et al.* 1991; Davis and Bortone, 1992; Bortone and Davis, 1994). Some contaminants, and particularly chlorinated organics, are known to function as synthetic hormones which disrupt normal endocrine and reproductive function (McLachlan *et al.*, 1992; Colborn *et al.*, 1993; Hileman, 1994).

Most of the biological effects observed in this river system are consistent with the findings of other studies that have investigated the impacts of pulp mill discharges on the health of fish populations. For example, a variety of biochemical, physiological, metabolic, behavioral and reproductive impairments due to exposure to pulp mill effluents have been reported in fish populations by Oikari and Nittyla (1985), Kovacs (1986), Andersson *et al.* (1987, 1988), McLeay (1987) and Owens (1991). Other pathological and reproductive effects have been documented by Couillard *et al.* (1988), Sandstrom *et al.* (1988), Lindesjoo and Thulin (1990) and McMaster *et al.* (1992). Effects caused by paper mill discharges have also been observed at the population and community levels (Karas *et al.*, 1991).

In summary, the biological effects observed in this aquatic ecosystem with the greatest human health implications appear to be those associated with:

1. genotoxic responses as evidenced by DNA damage;
2. reproductive impairment as seen in severely altered sex ratios and steroid hormone levels;
3. increased levels of cancer as demonstrated by the carcinomas in fish organs.

### 6.3.2 *ESTABLISHMENT OF CONTAMINANT PATHWAYS*

Once it has been established that a particular ecosystem has been impacted by contaminants and that the human population could also be at risk, the next step in developing this integrated approach is to identify potential links in terms of how contaminants are transferred from the aquatic system in question to the human population. Potential pathways which apply to this case are water contact, ingestion (drinking water and fish consumption), irrigation of crops and livestock watering.

Water contact occurs primarily from swimming activities. It is highly unlikely, however, that swimming is a health concern in this case given the general knowledge of the public concerning the pollution problems in this river. The ingestion pathway includes drinking water sources and fish consumption. Even though a large proportion of people that reside in the river valley utilize wells for drinking water, there is no evidence to date indicating that these wells are contaminated by river water. There is evidence, however, that fishing is a relatively common activity for some residents that live close to the river. Sport fishing for sunfish has been observed at several locations along the river and interviews with several residents have confirmed that fishing does occur on some regular basis. Whether these fish are consumed or not by the public has not been confirmed.

Livestock watering and crop irrigation are major activities occurring along various reaches of the river where toxicants have the potential to be transferred to the human population. Several dairy herds in western North Carolina are allowed to wade in and drink water from the river. In fact, several dairy barns are located within 50 m of the river where the cows have free access to the river bank. Since many of the chlorinated organic compounds such as dioxin are highly lipophilic and concentrate in fatty substances such as milk, the potential human health implications of this toxicant in the milk fat of dairy cows is obvious.

Irrigation of crops is a common activity along the river in the spring and summer. The major crop irrigated with water from the contaminated river is tomatoes with corn and tobacco receiving some irrigation during dry periods. Water is pumped directly from the river and sprayed on these crops along a 40 mile stretch of the river. In Cocke County, Tennessee, a 1,000 acre tomato farm is routinely irrigated with a series of large pumping stations that can deliver several thousands of gallons of water per hour. Pastures where cows graze are also occasionally irrigated with river water. Even though a contaminant such as dioxin is relatively hydrophobic and would probably not be incorporated into the plant through the root system, residues on the plants, fruits and livestock forage would be of concern relative to transmittal of pollutants to humans.

The major contaminant exposure pathway to the human population for this case appears to be livestock watering and crop irrigation. Both of these pathways have a potential to serve as a route of contaminant transmittal from the river system to humans. Water contact from drinking water and swimming or fish consumption do not, however, appear to be pathways of concern in this situation.

### *6.3.3 IDENTIFICATION OF BIOLOGICAL RECEPTORS*

Identifying common receptors and modes of action in the sentinel aquatic species and in humans is the third step in developing this integrated approach. Most of the pertinent information needed in this step is available from the literature. The principal target areas where the types of contaminants in this particular river generate the greatest concerns relative to human health are:

1. dysfunction of the endocrine and reproductive systems;
2. carcinogenic effects;
3. immune system impairment;
4. behavioral disorders.

Several recent studies on the effects of man-made chemicals on organisms have provided convincing evidence that many of these chemicals have the potential to disrupt the endocrine system of animals, including humans (Colborn *et al.*, 1993). The possibility exists that chronic low-level exposure to steroidogenic (estrogenic and androgenic) chemicals in the environment can have effects in humans similar to those observed in laboratory animals (vom Saal *et al.*, 1994). The lipophilic halogenated aromatic hydrocarbons such as PCBs and dioxin have been implicated in such endocrine-disrupting effects as thyroid dysfunction in birds and fish; decreased fertility in birds, fish, shellfish and mammals; gross birth deformities in birds and fish; masculinization and feminization of fish, birds and mammals; and compromised immune systems in birds and mammals (Colborn and Clement 1992). The mechanisms by which these chemicals have their impacts in animals vary, but they share the following general properties (Colborn and Clement, 1992).

1. Mimicking the effects of natural hormones by recognizing their binding sites.
2. Antagonizing the effect of these hormones by blocking their interaction with their physiological binding sites.

3. Reacting directly and indirectly with the hormone in question.
4. Altering the natural pattern of synthesis of hormones.
5. Altering hormone receptor levels.

Wildlife such as birds, fish and various mammals have provided the model for transfer of environmental endocrine-disrupting chemicals with their resulting suite of effects in offspring. Experiments with laboratory animals have confirmed these findings. It is difficult, however, to establish causal relationships between exposure to these xenobiotic compounds and human-related endocrine and reproductive effects (Reijnders and Brasseur, 1992). The next section presents information developed to establish possible causal relationships between contaminant effects on aquatic resources and human health effects.

### 6.3.4 *HUMAN HEALTH IMPLICATIONS*

In the final development phase of this approach, implications to human health are addressed in terms of providing evidence that suggests possible causal relationships between environmental and human health. In the absence of definitive epidemiological studies, which are not feasible in many cases, alternative lines of evidence can be used to build a case which suggests possible causal relationships between contaminant effects on aquatic resources and human health effects.

Statistical data for cancer incidence rates were obtained from the State of North Carolina, Central Cancer Registry (SCHES, 1992), and from the State of Tennessee, Department of Health (cancer reporting system). Data were reported for each of the counties in both states by race and sex and were age-adjusted using 10-year age intervals and the 1970 US population as the standard.

Of all 100 counties in North Carolina, Haywood County had the second highest incidence of female breast cancer (Figure 6.2). The pulp mill in question and the upper sections of the contaminated river are located within the boundaries of this western North Carolina county. Similarly, the county in eastern Tennessee where the downstream sections of the river are located ranks second of 22 east Tennessee counties for incidence of breast cancer (Figure 6.2). The probability, therefore, that these two counties through which the contaminated river flows would be simultaneously ranked both number 2 (in North Carolina) and number 2 (in east Tennessee) in breast cancer incidence by random chance is 2/1,100 (2/100 × 2/22). Therefore, there is only about a one in 500 probability that these observed incidences of breast cancer occur by random chance alone. This statistic does not, by any means, establish cause and effect but it is presented here as an example of how a case may be developed to address possible relationships between ecosystem and human health.

It is unlikely that the cancer incidence ranking for the counties in Figure 6.2

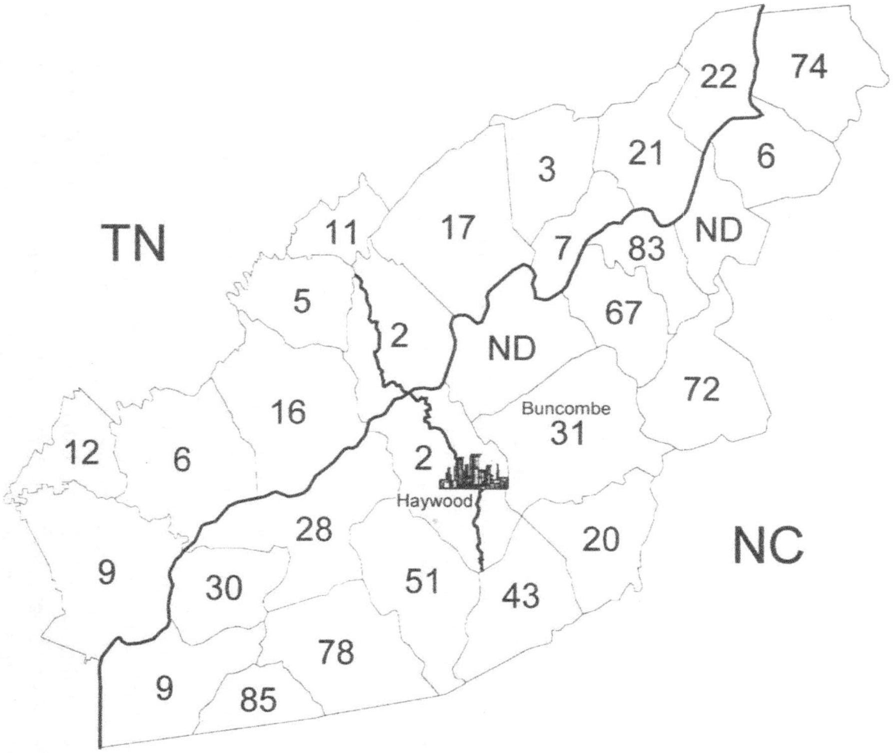


**Figure 6.2** Ranking of female breast cancer incidences in western North Carolina (NC) and eastern Tennessee (TN). Rankings for western North Carolina counties are based on all 100 North Carolina counties and ranking for east Tennessee are based on 22 east Tennessee counties. ND = no data available.

are strongly related to socioeconomic or other societal-based parameters because of the wide differences in county ranking within the same geographical area. For example, the western North Carolina counties shown in Figure 6.2 are all located within the Appalachian Mountain Province. It is assumed, therefore, that similar standards of living probably exist for people among those counties. This wide diversity in county ranking within a similar geographical area is particularly evident when Haywood County is compared to its surrounding counties. Even though Haywood County ranks second in the state for breast cancer incidence, its surrounding counties only rank 28, 51, 43, 20 and 31st, suggesting something unique about Haywood County relative to its higher incidence ranking. The only known major distinguishing characteristic in this county is the presence of the paper mill and the contaminated river which receives effluent discharges from this mill. Even Buncombe County, the highest population center in western North Carolina which is more industrialized than Haywood County, ranks only 31st in breast cancer incidence. To emphasize again, even though there appears to be a relationship between the location of the pulp mill and the incidences of breast cancer, this does not necessarily imply cause and effect but does indicate an area of concern where further studies are warranted.

This potential relationship between aquatic system and human health becomes even more convincing when we consider recent studies which show that endocrine (hormonal) disrupting compounds, such as dioxin which occur in this river, might be linked to breast cancer (Colburn *et al.*, 1993; Davis *et al.*, 1993; Wolff *et al.*, 1993). Estrogens are known to play critical roles in the development of breast cancer (Key and Pike, 1988) with the total cumulative exposure to bioavailable estrogens being the primary determinant (Henderson *et al.*, 1993). Approximately 40% of all cancers in women are believed to be hormonally mediated (Henderson *et al.*, 1991). Estrogens promote cell proliferation and hypertrophy of female secondary sex organs and induce the synthesis of cell type-specific proteins (Hertz, 1985). Some environmental chemicals that increase estrogen exposure by functioning as xenoestrogens may help explain, therefore, the sustained 1% annual increase in breast cancer mortality since the 1940s (Feuer and Wun, 1992; Davis *et al.*, 1993). A recent analysis of some chemical plant workers found more than a two-fold increase in breast cancer in females exposed to dioxin (Manz *et al.*, 1991). As discussed previously, some evidence of possible steroidogenic effects has been observed in fish populations from the upper sections of the contaminated river as demonstrated by the obvious impairments of normal sexual development (sex alterations and altered steroid hormone concentrations). Based on all the evidence available to date, it is reasonable to assume that these reproductive-related anomalies could be due to the steroidogenic effects of contaminants in this system. Deleterious effects on sexual development have been reported in a variety of wildlife species exposed to environmental contaminants (Colborn and Clement, 1992), such as those that occur in this study system.

## 6.4 SUMMARY

In summary, an approach using a four-step process has been presented that attempts to establish possible links and relationships between the health of aquatic systems and human health. Recognizing the limitations of such an approach in the absence of reliable epidemiological data, this model could be further tested and refined on other aquatic systems where environmental contamination is perceived to be a risk to human health. In addition, information generated by this type of study could also be used as an early warning signal of potential human health risks.

*ACKNOWLEDGMENT*

Oak Ridge National Laboratory is managed by Martin Marietta Energy Systems, Inc., under contract DE-AC05-840R21400 with the US Department of Energy.

## REFERENCES

Adams, S. M. (ed.) (1990) *Biological Indicators of Stress in Fish*, American Fisheries Society Symposium 8.

Adams, S. M., Crumby, W. D., Greeley, M. S. and Shugart, L. R. (1992) Responses of fish populations and communities to pulp mill effluents: a holistic assessment. *Ecotoxicol. Environ. Saf.*, **24,** 347–60.

Anderson, H. A. (1989) General population exposure to environmental concentrations of halogenated biphenyls, in *Halogenated Biphenyls, Terphenyls, Naphthalenes, Dibenzodioxins and Related Products* (eds R. D. Kimbrough and A. A. Jensen), Elsevier, New York, pp. 345–80.

Andersson, T. Bengtsson, B., Forlin, L. *et al.* (1987) Long-term effects of bleached kraft mill effluents on carbohydrate metabolism and hepatic xenobiotic biotransformation enzymes in fish. *Ecotoxicol. Environ. Saf.*, **13,** 53–60.

Andersson, T., Forlin, L., Hardig, J. and Larsson, A. (1988) Physiological disturbances in fish living in coastal water polluted with bleached kraft mill effluents. *Can. J. Fish. Aquat. Sci.*, **45,** 1525–36.

Blevins, R. D. (1991) 2,3,7,8-Tetrachlorodibenzodioxin in fish from the Pigeon River in Eastern Tennessee, USA: its toxicity and mutagenicity as revealed by the Ames Salmonella assay. *Arch. Environ. Contam. Toxicol.*, **30,** 366–70.

Bortone, S. A. and Davis, W. P. (1994) Fish intersexuality as indicator of environmental stress. *Bioscience*, **44,** 165–72.

Boudou, A. and Ribeyre, F. (1993) *Aquatic Ecotoxicology: Fundamental Concepts and Methodologies*, CRC Press, Boca Raton, FL.

Caruso, J. H., Suttkus, R. D. and Gunning, G. E. (1988) Abnormal expression of secondary sex characters in a population of *Anguilla rostrata* (Pices: Anguillidae) from a dark colored Florida stream. *Copea*, **4,** 1077–9.

Colborn, T. and Clement, C. (eds) (1992) *Chemically Induced Alterations in Sexual and Functional Development: the Wildlife/Human Connection (Advances in Modern Environmental Toxicology*, vol. 21), Princeton Scientific Publishing, Princeton, NJ.

Colborn, T., vom Saal, F. S. and Soto, A. M. (1993) Developmental effects of endocrine-disrupting chemicals in wildlife and humans. *Environ. Health Persp.*, **101,** 378–84.

Couillard, C. M., Berman, R. A. and Panisset, J. C. (1988) Histopathology of rainbow trout exposed to a bleached kraft pulp mill effluent. *Arch. Environ. Contam. Toxicol.*, **17,** 319–23.

Davis, W. P. and Bortone, S. A. (1992) Effects of kraft mill effluent on the sexuality of fishes: an environmental early warning? in *Chemically Induced Alterations in Sexual and Functional Development: the Wildlife/Human Connection* (*Advances in Modern Environmental Toxicology*, vol. 21) (eds T. Colborn and C. Clement), Princeton Scientific Publishing, Princeton, NJ, pp. 113–27.

Davis, D. L., Bradlow, H. L., Wolff, M. *et al.* (1993) Medical hypothesis: xenoestrogens as preventable causes of breast cancer. *Environ. Health Persp.*, **101,** 372–7.

Eisenreich, S. J., Looney, B. B. and Thornton, J. D. (1981) Airborne organic contaminants in the Great Lakes ecosystem. *Environ. Sci. Technol.*, **15,** 30–8.

EPA (1988) *Assessment of Dioxin Contamination of Water Sediment and Fish in the Pigeon River System.* US Environmental Protection Agency, Region IV. Report No. 001, Atlanta, CA.

Feuer, E. J. and Wun, L-M. (1992) How much of the recent rise in breast cancer can be explained by increases in mammography utilization? A dynamic model population approach. *Am. J. Epidemiol.*, **136,** 1423–36.

Gibbs, P. E., Spencer, B. E. and Pascoe, P. L. (1991) The American oyster drill, *Urosalpinx cinerea* (Gastropoda): evidence of decline in an imposex-affected population. *J. Mar. Biol. Assoc. UK*, **71,** 827–38.

Henderson, B. E., Ross, R. K. and Pike, M. C. (1991) Toward the primary prevention of cancer. *Science*, **254,** 1131–8.

Henderson, B. E., Ross, R. K. and Pike M. C. (1993) Hormonal chemoprevention of cancer in women. *Science*, **259,** 633–8.

Hertz, R. (1985) The estrogen problem–retrospect and prospect, in *Estrogens in the Environment II. Influences on Development* (ed. J. A. McLachlan), Elsevier, New York, pp. 1–11.

Hileman, B. (1994) Concerns broaden over chlorine and chlorinated hydrocarbons. *Chemical and Engineering News*, 19 April.

Kannan, N., Tanabe, S. and Tatsukawa, R. (1988) Potentially hazardous residues of non-orthochlorine substituted coplanar PCBs in human adipose tissue. *Arch. Environ. Health*, **43,** 11–14.

Karas, P., Neuman, E. and Sandstrom, O. (1991) Effects of a pulp mill effluent on the population dynamics of perch, *Perca fluviatilis*. *Can. J. Fish. Aquat. Sci.*, **48,** 28–34.

Key, T. J. and Pike, M. C. (1988) The role of estrogens and progestagens in the epidemiology and prevention of breast cancer. *Eur. J. Cancer Clin. Oncol.*, **24,** 29–43.

Kovacs, T. (1986) Effects of bleached kraft mill effluent on freshwater fish: a Canadian perspective. *Water Pollut. Res.*, **21,** 91–118.

Lindesjoo, E. and Thulin, J. (1990) Fin erosion of perch *Perca fluviatilis* and ruffe *Gymnocephalus cernua* in a pulp mill effluent area. *Dis. Aquat. Organisms*, **8,** 119–26.

Manz, A., Berger, J., Dwyer, J. H. *et al.* (1991) Cancer mortality among workers in a chemical plant contaminated with dioxin. *Lancet*, **338,** 959–64.

McCarthy, J. F. and Shugart, L. R. (eds) (1990) *Biomarkers of Environmental Contamination*, Lewis Publishers, Boca Raton, FL.

McLachlan, J. A., Newbold, R. R., Teng, C. T. and Korach, K. S. (1992) Environmental estrogens: orphan receptors and genetic imprinting, in *Chemically Induced Alterations in Sexual and Functional Development: the Wildlife/Human Connection*

(*Advances in Modern Environmental Toxicology*, vol. 21) (eds T. Colborn and C. Clement), Princeton Scientific Publishing, Princeton, NJ, pp. 107–12.

McLeay, D. J. (1987) *Aquatic Toxicology of Pulp and Paper Mill Effluent: a Review*. Report EPS 4/PF/1. Environment Canada, Ottawa, Ontario.

McMaster, M. E., Portt, C. B., Munkittrick, K. R. and Dixon, D. G. (1992) Milt characteristics, reproductive performance, and larval survival and development of white sucker exposed to bleached kraft mill effluent. *Ecotoxicol. Environ. Saf.*, **23,** 103–17.

Oikari, A. O. J. and Nittyla, J. (1985) Subacute physiological effects of bleached kraft mill effluent (BKME) on the liver of trout, *Salmo gairdneri. Ecotoxicol. Environ. Saf.*, **10,** 159–72.

Owens, J. W. (1991) The hazard assessment of pulp and paper effluents in the aquatic environment: a review. *Environ. Toxicol. Chem.*, **10,** 1511–40.

Rapaport, R. A., Urban, N. R., Capel, P. D. *et al.* (1985) New DDT inputs to North America: atmospheric deposition. *Chemosphere*, **14,** 1167–73.

Reijnders, P. J. H. and Brasseur, M. J. M. (1992) Xenobiotic induced hormonal and associated developmental disorders in marine organisms and related effects in humans: an overview, in *Chemically Induced Alterations in Sexual and Functional Development: the Wildlife/Human Connection (Advances in Modern Environmental Toxicology*, vol. 21) (eds T. Colborn and C. Clement), Princeton Scientific Publishing, Princeton, NJ, pp. 159–74.

Safe, S. (1987) PCB and human health, in *Polychlorinated Biphenyls (PCBs): Mammalian and Environmental Toxicology* (ed. S. Safe), Springer, Berlin, pp. 133–336.

Sandstrom, O., Neuman, E. and Karas, P. (1988) Effects of a bleached pulp mill effluent on growth and gonad function in Baltic coastal fish. *Water Sci. Tech.*, **20,** 107–18.

SCHES (State Center for Health and Environmental Statistics) (1992) *Cancer Incidence in North Carolina*. NC Department of Environment, Health and Natural Resources, Raleigh, NC.

Suter, G. W. (1993) *Ecological Risk Analysis*, Lewis, Boca Raton, FL.

vom Saal, F. S., Finch, C. E. and Nelson, J. F. (1994) Natural history and mechanisms of aging in humans, laboratory rodents, and other selected vertebrates, in *Physiology of Reproduction* (eds E. Knobil, J. Neill and D. Pfaff), Raven Press, New York, pp. 1213–1314.

Wolff, M. S., Toniolo, P., Lee, E. *et al.* (1993) Blood levels of organochlorine residues and the risk of breast cancer. *J. Natl Cancer Inst.*, **85,** 648–52.

# PART THREE

## *Interdependence between Human and Ecosystem Health*

# 7 Ecosystems as buffers to human health

FUMIO MATSUMURA

## 7.1 INTRODUCTION

Many studies have been conducted on the effect of man-made pollution on ecosystems and the effects of environmental pollution on human health. Understandably, biological effects of toxic pollutants are of great concern to many segments of any society and particularly those dealing with the effects of pollutants on human and ecological systems.

However, the topic of how ecosystems affect human health is not an easy subject to deal with. First, even when a clear-cut case of pollution in a local area is discussed, the effect of ecosystems on the eventual toxic interactions of the polluting materials with the human population is indirect, secondary at best, and in most parts undefinable. Furthermore, one may always struggle with the definition of ecosystems and argue what constitutes 'healthy' ecosystems.

The approach I have chosen here in preparing this chapter is to address this topic from the specific viewpoint of the 'buffering influences' of ecosystems on relatively well-defined environmental health problems affecting human populations. The term 'ecosystems' is interpreted as a broad and loose entity containing more than one species of wildlife or even man-managed systems (e.g. agroecosystems) in a given area.

'Buffering' effects occur when the given effect of toxicant is somehow reduced, delayed (e.g. the same dose of chemical spread over a long time period) or qualitatively altered in the presence of the buffering factor, ecosystems in this case, as compared to the same situation without the same factors. Such buffering actions could be produced when the given ecosystems (or their components) act as shields, sequesters, surrogates, metabolic convert-

*Interconnections Between Human and Ecosystem Health.* Edited by Richard T. Di Giulio and Emily Monosson. Published in 1996 by Chapman & Hall, London. ISBN 0 412 62400 1.

ers, diluters, competitors, translocators, filters, temporary resistance, physical barriers, providers of antagonists, diverters or helpers in healing, etc. It must be pointed out that not all ecosystems act as buffers and in certain cases ecosystems themselves could produce toxic or pathogenic factors or synergize and could amplify environmental health problems (e.g. plant pollens could exacerbate asthma conditions). Therefore, the cases mentioned in this chapter are mainly those which are clearly beneficial in nature from the human's point of view. The antagonistic cases will be mentioned also to provide the reader with the chance to gain some perspectives.

## 7.2 TRENDS IN HUMAN ENVIRONMENTAL HEALTH SCIENCES

Environmental health science is a field of medical science specifically addressing environmental factors affecting human health. Indeed many factors are known to significantly influence human health (see Trieff 1980; Lippmann and Lioy, 1985; Moeller, 1992). The most powerful tool in this field is epidemiology which is used in detecting positive correlations between environmental factors and health problems. A quick glance at trends in environmental health problems in the US in the past 100 years reveals several definite changes (e.g. Feinleib and Wilson, 1985). The most noticeable change is the rise in life expectancy. In 1900 the life expectancy of white males and females was in the order of 50 and that for all other races was estimated to be only 34 years. Today, the life expectancy of average US citizens has risen to 78 for females and 73 for males. Among causes of death, the decline of strokes has been particularly noticeable (about 50% since the 1960s), although that of coronary heart disease (30% decline in the same period) is also substantial. While the leading cause of death is still combined arterial diseases (874 000 per year), the fastest rising category is death from all types of cancer (330 000), surpassing all accidental deaths (115 000) to become the second leading cause. Another fast rising category is combined respiratory diseases (113 000). Experts in environmental health sciences are in general agreement that environmental contributions to the genesis of cancer and respiratory diseases are quite significant, though there are varying degrees of their extents of influence as compared to other causes such as genetics, abuse, etc.

With regard to non-fatal diseases (collectively termed 'morbidity'), the frequency of chronic activity-limiting diseases is increasing (during the period between the 1960s and 1980s the average rate jumped from 5% to approximately 10% of the population). The major causes among women aged 45–64 are arthritis and rheumatism 24.2% and heart conditions and hypertension 22.2%, whereas the corresponding figures for males (all ages) are heart conditions 16.7% and arthritis and rheumatism 11.4%. Other major chronic diseases causing frequent debilitation are pneumonia and diabetes. While the rate of the former has stayed relatively constant, that of the latter has steadily increased (approximately 2.2% in 1960 to 5% of the US population, i.e. over 100% increase).

Among cancer incidences, the most spectacular rise is still seen in the category of lung cancer (60% to 150% increase from 1973 to 1987 for age less than 65 and over 65, both males and females respectively, Marshall, 1991). Others showing appreciable increases are (in order of decreasing importance) melanoma, multiple myeloma, esophagus, renal, liver, non-Hodgkin's lymphoma, prostate, breast and brain (only the over 65 age group). Decreasing trends have been observed in cancers of testis, cervix, uterus, bladder, stomach and thyroid and oral cancers.

These data indicate that air pollution is the major factor in increasing lung diseases. Other environmental factors affecting human health are hormone mimics as judged by the increase in breast and prostate cancer and decreasing sperm counts. Another area requiring serious attention would be the increase in immune-related problems such as the rise in multiple myeloma, leukemia and autoimmune diseases (arthritis, Type I diabetes, etc.). The overall patterns and the balance of evidence support the view that man-made activities indicating epidemiologically positive correlations should be areas for future studies.

A particularly important question relative to consideration of ecosystems as buffers to human health is whether there is any evidence that a non-urbanized environment with a minimum of human activity and an abundance of wildlife provides a healthier environment for humans than a metropolitan area. Perhaps the simplest example is in the case of lung cancer incidence. Expressed in cases per 100 000, the incidence is 52 for those living in areas with a population over 50 000 as compared to 44 for people living in areas with a population between 10 000 and 50 000. The values for suburban dwellers and rural residents are 43 and 39, respectively (all corrected and adjusted for age and cigarette smoking) and are statistically different from each other, indicating clearly that the closer one comes to a metropolitan area, the higher the chance of developing lung cancer (Hammond and Horn, 1958).

## 7.3 AIR POLLUTION

### *7.3.1 CARBON DIOXIDE*

The atmospheric concentration of carbon dioxide has increased dramatically in recent years (e.g. Graedel and Crutzen, 1990). The current level, 350 000 ppb, represents about a 20% increase in the past 100 years. Carbon dioxide has a long average residence time of 100 years in the atmosphere, and appears to be well correlated to the past record of temperature changes of the earth over a 150 000 year time span estimated from Antarctic ice records. The so-called 'greenhouse' effect named in anticipation of future global warming is considered to be largely due to an increasing $CO_2$ level in atmosphere. While there are discussions and arguments over the possible extents of global warming according to the expected rise in $CO_2$, most experts appear to agree that warming is a definite trend (Schneider, 1990). Added to the above problem is the expected rise in demand for energy, foods and further industrialization as a result of the future population growth which is projected to increase at a rate of 1 billion every 10 years.

According to Schneider (1990) the atmospheric quantity of $CO_2$ is estimated to be in the order of 740 billion metric tons with an annual increase of 3 billion tons. The estimated anthropogenic inputs are 5 billion tons through fossil-fuel utilization and 1–2 billion tons through deforestation/burning activities. The ocean is a net importer of $CO_2$ by 3 billion tons and the entire terrestrial system is currently balanced. The key figures in this balancing act in the total terrestrial environment are outputs of 54–55 billion tons from soil, 55 billion tons from plant respiration and 110 billion tons of $CO_2$ reclaimed by plants during photosynthesis. Thus, healthy plants carrying out rigorous photosynthesis are the major sequesters of $CO_2$, providing a powerful 'buffering' effect in this balancing act of global $CO_2$ output and input equation. Unfortunately, deforestation is proceeding rapidly and significant desertification is occurring in some parts of the world.

### 7.3.2 *OZONE AND SULFUR DIOXIDE*

While the effect of $CO_2$ is important to human health from a global point of view, there are much more reactive, direct poisons in the atmosphere causing damage to the respiratory system including the nasal regions of man. Both $O_3$ and $SO_2$ are such poisons closely associated with air pollution. $SO_2$ is much more water soluble than $O_3$ and therefore the former causes damage at the nasal to the upper tracheal region and the latter tracheal to upper lung region. In addition, $SO_2$ attached to small particles less than 3 μm in diameter could be carried deep into the lung. Both acute (edematosis) and chronic problems (emphysema, lung cancer, etc.) have been linked to exposure to these air pollutants.

Interestingly, many plants are known to absorb these pollutants and some are severely damaged by them (Feder, 1978). These include wide-ranging plants such as pines and other coniferous trees. Another group of organisms known to collect and accumulate these air pollutants are lichens and mosses (eukaryote plants). In fact, some species such as *Hypnum cupressiforme* moss have been used as monitoring tools for sulfur dioxide, as well as heavy metal residues coming from air pollution (Feder, 1978). Some are also sensitive to sulfur dioxides, as it has been noted that there are some 'lichen deserts' and 'moss deserts' in the vicinity of sources of sulfur dioxide. In all cases, however, there appear to be tremendous species sensitivity differences and, at moderate pollution level, certain species tend to thrive, indicating that the availability of a diverse spectra of species provides the buffering capacity of ecosystems.

Unfortunately, there are no quantitative data illustrating exactly how much $O_3$ and $SO_2$ are being removed by plants. On the other hand, the key health consideration is the period of peak concentration of some of these pollutants, when levels exceed health standards within the breathing space for the residents. The concentration of $O_3$, for instance, hits its peak in the afternoon during summer months when sunlight intensity peaks. Any factor which can reduce, delay or dilute this concentration would be of great help.

### 7.3.3 *NOx PARTICULARLY $NO_2$ POLLUTION*

$NO_2$ is produced by incomplete combustion processes (i.e. the original source of nitrogen is atmospheric $N_2$), and therefore mostly associated with air pollution around big metropolitan cities with high automobile traffic activity. Among these air pollutants, $NO_2$ is the most serious health hazard, since its water solubility property is such that this gas penetrates into the deep lung region where it causes the most damage. $NO_2$ is also an important trigger for ozone formation in sunlight.

$$NO_2 \xrightarrow{\text{light}} NO + (0)$$

$$0_2 + (0) \longrightarrow O_3$$

Unfortunately, $NO_2$ is not taken up well by rain or absorbed into fog and other moisture-containing materials. Nor does it appear that soil microorganisms can reduce atmospheric $NO_2$ (as a part of nitrification or denitrification) in sufficient quantities to affect the total atmospheric $NO_2$ at peak pollution time.

However, under strong sunlight both $NO_2$ and $O_3$ are known to react with short chain hydrocarbons.

$$CxHy + 0_3 + NO \xrightarrow{\text{light}} CxHy\ O_3\ NO$$

Thus, it is likely that such reactions take place at the surface of plants as well as in the atmosphere with short chain alkyl compounds emitted by plants.

Again, no quantitative data are available to indicate how much of a buffering effect plants offer in any of the air pollution regions.

### 7.3.4 *HEAVY METALS AND RADIONUCLIDES*

Heavy metals cause a variety of health problems. Cadmium induces kidney tubular damage, lead is associated with nervous system damage, IQ deficit, gastrointestinal distress, anemia and respiratory problems, methyl mercury damages the central nervous system, arsenicals produce mitochondrial and nerve damage as well as cancer and one form of nickel induces cancer, etc. It must be mentioned here that there are several routes for man's exposure to heavy metals. Oral uptake through food and drinking water is an important route. Some of them, such as lead and mercury, are significant air pollutants and lung toxicants. Radon is an excellent example of a toxic metal in the form of a gas. Its human toxicity is known to be mainly due to its direct lung toxicity rather than generalized radiological tissue damages.

Two aspects of metal toxicology are very pertinent to this discussion. One is the toxicologically relevant form of metals (metal species) and the other is their general propensity to bind with the SH-moiety of cysteine, glutathione, etc. present in many biological systems. It is noteworthy that many organisms produce metallothionein type proteins that are rich in cysteine residues, and apparently specialized for the protection of species from metal toxicity. Thus the large-scale availability of biological fauna and flora in the vicinity of the pollution source is likely to provide sequestering masses to buffer the direct exposure of man to heavy metals and those radionuclides possessing SH-binding properties. Indeed, high metal and radionuclide concentrations have been found in many plant species (Feder, 1978; Koranda and Robison, 1978), and aquatic organisms such as fish and terrestrial animals (Pacyna and Ottar, 1989). A key point to remember is that such bioconcentrating phenomena could also adversely affect man as shown in the case of human poisoning incidence of cadmium (rice, Japan), methyl mercury (fish, Japan), $^{131}I$ (milk from cows consuming grass, Soviet Union after a nuclear reactor accident), etc. The important factor in deciding the ratio of beneficial versus adverse roles of ecosystems would be the biodiluting versus bioconcentrating properties of their biological components.

Another salient point is that metal toxicity could vary greatly both qualitatively and quantitatively according to their forms. In the case of arsenicals, $AsO_4^{-3}$ and $AsO_3^{-3}$ are the most toxic forms. In nature, a series of methylation processes are known to take place. In most cases these biological, metabolic transformation processes are regarded as detoxification reactions. Thus the highly metabolized forms such as $(CH_3)_3A + s\ CH_2\,COO^-$ (betaine) found in shrimps and other aquatic invertebrates are considered as practically non-toxic. Accordingly in this case, robust biological activities clearly serve as buffering processes to metal toxicity. On the other hand, in the case of mercury, the most toxic form is methyl mercury, a microbial metabolic conversion product from $Hg^{2+}$ . The reaction occurring at a relatively narrow redox potential range in aquatic sediments appears to be responsible for its formation in the environment. Thus not every biological input results in beneficial outcomes. On the other hand, even methyl mercury accumulation in fish in aquatic systems is strongly affected by competing biological systems and pH (i.e. biological accumulation is very high in pH lower than 5). Therefore, mercury-related health problems are seldom seen in areas rich with biological activity, such as eutrophic lakes and estuaries, since they provide high titers of -SH sources as well as a neutralizing influence on acidic water.

## 7.4 PESTICIDES AND TOXIC ORGANIC CHEMICALS

The level of annual production of pesticides in the US was around 500 000 tonnes in the early 1970s. While it has declined to the current level of approximately 230 000 tonnes per year still colossal amounts of pesticides are used today in this country. Yet, the bulk of those pesticides, particularly modern organic pesticides with relatively short residual properties, are known to dissipate in agroecosystems within one annual cleansing cycle. Certainly, some of the dissipation is due to dilution, dispersion, translocation and physical or chemical conversion and sequestration. However, in terms of overall quantitative balance, metabolic degradation in the environment appears to be the most significant cleansing force eliminating their residues. Since this particular topic is well covered (e.g. Matsumura and Krishna Murti, 1982; Grisham, 1986; Bourke *et al.*, 1992), there is no need to go into the details of pesticide-degrading powers of soil microbial ecosystems. The only point requiring re-emphasis is the fact that pesticide wastes piled up in dry landfills or any other non-microbially active sites are not going to disappear. Even in microbially active environments such as agroecosystems and forest floors, the actual rates of degradation can vary widely. However, the overall picture is clear that microbial degradation of pesticides proceeds well in areas with rich microbial fauna and flora. Science in this field has progressed to the extent that today general microbial degradation potential for each pesticidal chemical in given environments may be predicted with good levels of confidence. However, heavy uses of soil fungicides/fumigants could wipe out those useful pesticide-degrading fauna and flora as well as pest insect pathogens and parasites. Well-thought-out and well-orchestrated pest management/sustainable agriculture approaches are the way to reduce pesticides while avoiding the development of resistance among pests.

## 7.5 ENVIRONMENTAL PATHOGENS AND INFECTIOUS DISEASES

It is often mentioned that pathogenic microorganisms are typically associated with polluted waters (e.g. Sykes and Skinner, 1971). This is indeed the case with large numbers of pathogens enteric in origin being discharged via sewage effluents. A contamination case of municipal drinking water in Milwaukee in the summer of 1993 illustrates such an example. Among those pathogens, enteroviruses, salmonellae and *E. coli* are probably most important in the US, but others such as *Vibrio cholera*, *Shigella* and *Entamoeba histolytica* may be very serious in other parts of the world.

Although water contamination cases are known to occur occasionally, these pathogenic organisms generally do not persist long in natural waters (Hawks, 1971). This fortunate phenomenon is not usually due to simple dilution. Rather, these organisms are eliminated or reduced by competition, predation and inhospitable environmental factors such as sunlight, high oxygen contents and the lack of nutrition. Alternatively, they tend to survive in waters with high concentrations of sewage effluent. Also important is the consideration that some of these pathogenic organisms (e.g. enteric viruses, helminth and bacteria) could persist long in shellfish living in polluted water even after they have disappeared from the water. Thus from the viewpoint of human health the presence of edible shellfish could exacerbate the problem, though the presence of healthy competitive microbial ecosystems would undoubtedly lessen the population and the impacts of pathogenic organisms.

The key to the maintenance of pathogenic organism-free (i.e. low enough to cause no health problems) waters appears to be maintaining low nutrient, high oxygen, and good sunlit aquatic ecosystems. The presence of natural filtration systems such as wetlands are known to greatly reduce enteric bacterial counts entering open water.

As for insect-borne infectious diseases, the past four decades have witnessed a tremendous progress in reducing major diseases such as malaria, yellow fever, filariae, etc. Certainly these are the results of wise use of pesticides and elimination of the habitats for the vector species and wild hosts. On the other hand, malaria is showing some sign of a comeback and some major vectors (e.g. tse-tse flies) still resist eradication efforts by man. Establishment of pest management strategies with the goal of establishing stable ecosystems including predators (e.g. mosquito fish, predatious mosquito species), parasites (e.g. nematodes), sympatric competitor species (that are not potential vectors), and other competitors to nutrition and space are the desired approach. Although these may not be considered as natural ecosystems from the viewpoint of wildlife ecologists, these approaches represent the wise selection of utilization of the biological diversities offered by nature to achieve a balanced existence of man and wildlife.

## 7.6 CONCLUSION

The natural environment serves as an effective buffer to pollution and pathogenic microorganisms. Certainly, some of the buffering actions are due to simple physical dilution effect. Nevertheless, there are a number of concrete cases where various components of ecosystems have been shown to play strong roles in reducing the hazards of pollutants and pathogenic organisms by actively eliminating, sequestering, quenching, acting as alternative targets, reacting, metabolizing, competing and providing physical barriers, traps and filters, etc. From the viewpoint of air pollution, plants play by far the most significant role because of their biomass, their combined large surface areas, as well as their unique ability to fix $CO_2$ and other air pollutants. From the veiwpoint of the total pollutant-cleansing capacity of this earth, soil microorganisms perhaps constitute the most significant factor. Most pesticides and other organic pollutants in the environment are degraded by them. The other side of the coin is that in areas where microorganisms cannot thrive, such as in dry landfills, deserts, Arctic and Antarctic regions, superoligotrophic lakes, etc., these pollutants could continue to accumulate.

From the viewpoint of environmental health scientists, the fastest rising environmental problems in the industrialized world are lung diseases, autoimmune diseases (diabetes, arthritis and rheumatism) and hormone-related afflictions (increases in breast cancer, prostate cancer, a decrease in sperm counts, etc.). While in many cases we do not know the precise cause, these are the problems less frequently observed among people in less industrialized and urbanized environments or among past generations living before the post Second World War period. While the precise statistics and the concrete proofs for the beneficial roles of health ecosystems are difficult to find, it is clear that in terms of overall balance, many biological components of ecosystems contribute in reducing the total toxic effects of man-made pollutants. Careless elimination of viable ecosystems such as forests, wetlands, grasslands and water bodies, or overcoming their capacities to handle toxic insults, would decrease the overall cleansing and buffering capacity of the environment and eventually affect man's health in adverse ways.

## REFERENCES


Bourke, J. B., Felsot, A. S., Gilding, T. J. *et al.* (eds) (1992) *Pesticide Waste Management. Technology and Regulation*, ACS Symposium Series 510, *Am. Chem. Soc.*, Washington, DC, pp. 1–273.

Feder, W. A. (1978) Plants as bioassay systems for monitoring atmospheric pollutants. *Environ. Health Perspect.*, **27**, 139–47.

Feinleib, M. and Wilson, R. W. (1985) Trends in health in the United States. *Environ. Health Perspect.*, **62**, 267–76.

Graedel, T. E. and Crutzen, P. J. (1990) The changing atmosphere, in *Managing Planet Earth* (Board of Editors *Scientific American*), W. H. Freeman, New York, pp. 13–24.

Grisham, J. W. (ed.) (1986) *Health Aspects of the Disposal of Waste Chemicals*, Pergamon Press, New York, pp. 1–454.

Hammond, E. C. and Horn, D. (1958) Smoking and death rates – report on 44 months of follow-up of 187,783 men. *J. Am. Med. Assoc.*, **166**, 1159–72, 1294–1308.

Hawks, H. A. (1971) Disposal by dilution – an ecologist's view, in *Microbial Aspects of Pollution*, (eds G. Sykes and F. A. Skinner), Academic Press, San Diego, CA, pp. 149–79.

Koranda, J. K. and Robison, W. L. (1978) Accumulation of radionuclides by plants as a monitoring system. *Environ. Health Perspect.*, **27**, 165–79.

Lippmann, M. and Lioy, P. J. (1985) Critical issues in air pollution epidemiology. *Environ. Health Perspect.*, **62**, 243–58.

Marshall, E. (1991) Experts dash over cancer data. *Science*, **250**, 900–3.

Matsumura, F. and Krishna Murti (eds) (1982) *Biodegradation of Pesticides*, Plenum Press, New York, pp. 1–312.

Moeller, D. W. (1992) *Environmental Health*, Harvard University Press, Cambridge, MA, pp. 1–332.

Pacyna, J. M. and Ottar, B. (eds) (1989) *Control and Fate of Atmospheric Trace Metals* (NATO AS1 series. Series C: *Mathematical and Physical Sciences*, vol. 268), Kluwer Academic Press, Dordrecht, pp. 1–382.

Schneider, S. H. (1990) The changing climate, in *Managing Planet Earth* (Board of Editors *Scientific American*), W. H. Freeman, New York, pp. 25–36.

Sykes, G. and Skinner, F. A. (eds) (1971) *Microbial Aspects of Pollution*, Academic Press, San Diego, CA, pp. 1–289.

Trieff, N. M. (1980) *Environment and Health*, Ann Arbor Science Publishers, Ann Arbor, MI, pp. 1–652.

# 8 Interfacing product life cycles and ecological assimilative capacity

JOHN CAIRNS, JR

## 8.1 DEVELOPING A PARTNERSHIP BETWEEN TECHNOLOGICAL AND NATURAL SYSTEMS

The planet could not support the present human population of just under 6 billion without the technological life support system that has been evolving since the agricultural revolution began. The average citizen recognizes a dependence upon components of the planet's technological life support system (such as fossil fuel energy, the transportation system, the communication system, and the like) even without understanding how they function.

In the past, the planetary biosphere was one vast interactive system. Virtually everything that the planet produced was part of the natural recycling system. This is why fossil remains of early creatures are so extraordinarily difficult to find despite the fact that numerous species were almost certainly abundant. Technological artifacts from early civilizations are relatively rare since they do not cover large geological epochs. Some early civilizations did, of course, produce enduring artifacts such as the Great Wall of China, the Pyramids of Egypt, and the Aztec and Mayan ruins in the Americas of fairly sophisticated civilizations. The fact that even these cover a minuscule percentage of the earth's surface demonstrates the effectiveness of natural recycling processes in earlier times.

However, the present technological civilization has produced fairly substantial areas devoted to highways, cities, irrigation canals and the like. Again, these cover only a relatively small portion of the land mass of the earth, but they do represent a far, far higher percentage of the total land mass than the artifacts of earlier civilizations. Some less visible but perhaps even more serious problems exist with storage of long-term radioactive wastes, persistent hazard-

*Interconnections Between Human and Ecosystem Health.* Edited by Richard T. Di Giulio and Emily Monosson. Published in 1996 by Chapman & Hall, London. ISBN 0 412 62400 1.

ous wastes and storage of the municipal solid waste (MSW) of today's societies buried in landfills which, although less hazardous than the other wastes just mentioned, far exceed them in volume. Some of these storage sites, arguably eventually all, form leachates that may enter ground water or surface water and sometimes may appear on the surface. Since modern landfill practices in the US require the use of daily cover with a layer of soil or sand, this may be less of a concern in the future. That these wastes all have potential for adversely affecting human health seems fairly conclusive. There is, naturally, heated debate about the degree of risk or hazard. Relatively few landfills achieve zero risk from these waste materials that have not been reincorporated into natural cycles.

Probably, pesticides and radioactive contaminants first demonstrated persuasively to the general public that materials in the environment that had deleterious effects on other organisms could and did have disastrous effects on humans. In some cases, contaminants reached humans primarily through the food they ate. Acute effects on human health are powerful demonstrations of risks not well understood by the general public. They persuaded people that a relationship did exist between human health and environmental conditions, but, when the drastic symptoms disappeared, it was possible for most citizens to believe that the problem had disappeared as well. People turned their attention to some more immediate crisis. The result is a short attention span on major problems for which the consequences are clearly evident. In contrast, this discussion focuses on much more subtle relationships, which many humans do not take seriously.

## 8.2 THE FRAGILITY OF NATURE MYTH

Since the first Earth day over two decades ago, the public has been bombarded with stories about how fragile and easily disturbed natural systems are. This is, of course, quite true because ecosystems such as tropical rain forests, tundra and a variety of others are not easily restored once damaged. In fact, precisely replicating a damaged ecosystem may be impossible, no matter how elegant the restoration efforts (Cairns, 1989).

However, another perspective deserves more attention: while the structure and function of complex ecosystems may be altered either deliberately or inadvertently, some natural systems are tough and resilient. The attention focused on the fragility of some species has led to a mind set in much of human society that assumes all species survive only at mankind's discretion. It is extremely difficult and arguably impossible to completely eliminate all living things. Life exists, even flourishes, in the thermal vents on the ocean floors, under extraordinary conditions thought impossible for living things before the discovery (Tunnicliffe, 1992). The difficulty human society has had in modifying nature can be seen through many examples. Human society has tried, thus far unsuccessfully, to keep the fruit fly out of California. The difficulties of beating back kudzu vines and other exotics deliberately introduced into this country are legendary. Massive attempts have been set in motion to eradicate various pests affecting agricultural crops, the gypsy moth in eastern forests, and the Asiatic clam and zebra mussel that are widely spread throughout the United States (Office of Technology Assessment, 1993).

Human society must develop a partnership with nature, which, at the very least, requires that mankind not exceed the non-degrading assimilative capacity of natural systems for societal wastes and, at the very best, means that mankind remain below the assimilative capacity threshold to a point where natural systems thrive.

If ecosystems include tough species why worry about exceeding assimilative capacity? The major reason is that these tough species are the species that humans cannot control; in other words, pests. Not exceeding natural assimilative capacity will permit the less obnoxious species that provide benefits that exceed costs to survive. If human society continues on its present course, the species that will survive and thrive will be the ones most tolerant of the conditions produced by human society, and they will probably be extremely difficult to control.

Some thought should be given to the development of a partnership with nature that ensures continuation of the many services (such as maintaining the atmospheric gas balance, maintaining water quality) that natural systems provide to human society automatically and at little or no cost.

## 8.3 RECYCLING AS A DESIGN FUNCTION FOR NEW PRODUCTS

If human society seriously attempts to mimic the conditions that preceded the agricultural and industrial revolutions (co-existing with rather than dominating natural systems), sustainable use of natural systems is more likely. Conditions in which ecological resources were temporarily used by various species (including humans) and quickly, almost gracefully, reincorporated into the biosphere are the basis of sustainable use. If sustainable use is a goal, no alternative remains but to consider mimicking natural, cyclical processes as a component of product development. This will require major behavioral changes at the individual level and at various organizational levels of society, including those of industry and government.

This paradigm shift should occur first in the educational system because this is the area in which students either develop a multidimensional view of the discipline or learn to view their field as isolated from the other disciplines. Too much emphasis on specialization decreases interactions with other disciplines, which in turn diminishes any effectiveness in coping with problems that transcend the capabilities of a single discipline (i.e. all of human society's major problems).

Holl *et al.* (1995) provide an illustration of how simple design considerations can improve the compatibility of commonly disposed products with recycling systems. Small design differences can make a product relatively easy for a recycler to process or make it virtually impossible to do so. As an example, polyethylene teraphthalate (PET) and polyvinyl chloride (PVC) are both common polymers used in packaging as well as in a variety of other applications. Each is technically easy to recycle. However, in combination, these two polymers are a disaster since PVC melts at 100°C while PET does so at 255°C. Therefore, PET remains solid in PVC recycling systems and plugs up extrusion screens and orifices. In contrast, very small amounts of PVC in a PET recycling system will melt, char, produce chlorine gas and drastically reduce the viscosity of the molten PET so that it is unprocessable. This is perhaps the most dramatic example of the importance of design for recyclability, although a number of similar instances can be found (Holl *et al.*, 1995). It could be argued that recyclability was not a part of the design criteria of PET or PVC because they were invented years before recycling became an issue. To the contrary, had recyclability been considered *before* the products were marketed, recycling would not have become an issue. Is it a sin to avoid problems by viewing everything in a multidimensional context?

## 8.4 CULTURAL EVOLUTION

Almost everything in the biological world can be transformed by some species. For example, the bones in a skeletal support system can be transformed by other creatures when its original owner dies into some other structure of temporary use to another individual, probably in another species. However, the types of materials produced by a technological society, such as plastics, computer chips, persistent toxicants and discarded automobiles, are not as readily reincorporated into natural systems. The evolutionary process has simply not had long enough to develop transforming species and, for many materials, appears unlikely to do so, especially at the current rates of production. Therefore, human society will either have to produce less of these materials or assist natural systems in the reincorporation process.

Natural recycling can be mimicked in two ways: first, by continually reusing materials extracted from natural systems so that the recycling occurs entirely within the technological system or, secondly, by designing most of human society's artifacts so that they are readily assimilated by natural systems. In both cases, the process of recycling must be considered in the design phase of a new product so as to avoid accumulations of society's wastes and discards in landfills and other burial grounds. In places, designing with recycling in mind will require long-term planning and the collaborative efforts of various elements of society such as industrialists and environmentalists. Recycling by design must have both a 'bottom-up' strategy, beginning with the individual, and a 'top-down' strategy, integrating activities of the various organizational components of society, including the educational system, industries, and government (Holl *et al.*, 1995). Waste minimization is an obvious key element at both the individual and organizational levels. Waste minimization would effectively reduce the magnitude of the problem almost immediately. With this effort, recycling can be considered first within the technological system by determining how the product can be made more amenable to this end. The word *product* is used here in a very broad, general sense to include all parts of the process necessary to produce the item, not just the item itself.

Re-introducing or recycling products of a technological society into the biosphere will require the emergence of a reverence and respect for life and the environment. Diamond (1992) examines this in his book *The Third Chimpanzee* and, in a much more condensed form, in the *Bulletin of the American Academy of Arts and Sciences* (Diamond, 1994). However, Diamond examines a more disquieting possibility: earlier civilizations damaged the environment as much as the technological means available to them permitted. While there were undoubtedly tribes that had a reverence for the interdependent web of life called ecosystems, such 'environmentalists' may have been no more common than they are today. Extinctions of various species, such as song birds, etc., did not necessarily occur because they were used for food or because their habitats were destroyed, but rather because predators or other species introduced by humans effectively wiped them out. If Diamond's speculations are correct (and his analysis is certainly persuasive), then the current society has the opportunity to be the first civilization to reach a true partnership with nature on a large scale.

The major advantage that current societies have over earlier ones in pursuing this goal is the existence of the information age, including computers that cope with vast amounts of information. However, this information will not be used in an optimal way unless accompanied by major societal change. As Holl *et al.* (1995) note, making better use of ecological and technological resources requires certain basic changes in current thought patterns. First, *garbage* must be redefined. The word *garbage* means something useless or of no value. While society commonly considers aluminum cans as garbage, they have much value, as does organic waste, paper and metals other than aluminum. When garbage is redefined, it is apparent that the majority of what is currently thrown away may be usable in a variety of ways. Second, people must reconsider society-acceptable practices. Most people cringe when they see someone throw trash from a car window, but they do not think twice about using single-use cups or throwing an aluminum can in the garbage. In many European countries, shopping bags are not provided by stores, so shoppers have learned to bring reusable bags when shopping. Third, the costs of products must be re-evaluated. The price consumers pay for products must not only reflect the cost of production, but also the environmental costs of extracting the necessary materials, the disposal of toxic byproducts, the energy required to manufacture the product, the transport costs and the ultimate cost of disposal or recycling the product. Fourth, people must begin to consider the long-term effects of their daily actions. While many consumption and disposal decisions seem trivial at the individual level, these, in the aggregate, almost invariably have a major impact. Fifth, communities must dispose of their wastes locally in so far as is possible. This means that the whole process can be viewed by the society, rather than the present process of fragmenting and hiding parts of the disposal process. Society would be much less likely to produce hazardous waste sites, or even unsightly waste disposal sites, if they were in its own backyard.

## 8.5 THE EDUCATIONAL SYSTEM: COMPLICATED VERSUS COMPLEX

Jacques Barzun (1964), in *Science: the Glorious Entertainment*, makes a distinction between 'complicated' and 'complex'. In the Second World War many people were taught to dissemble and reassemble complicated equipment, sometimes in the dark. Old-fashioned, spring-driven watches were complicated, but assembly could be accomplished by rote learning. Human relationships and society, on the other hand, are complex and their problems cannot be solved by a rote approach. Technology and 'big' science and engineering have produced a situation that has led comprehensive research universities to stress the complicated rather than complex. The epitome of this view is to be found in a statement by Mary Lowe Good, chairperson of the National Science Board:

> University science and engineering courses designed to weed out all but the most able need to be redesigned. The advice now must be, 'Look at the client in the house and ask what I can do to make this a useful student.' Instructors must ask, 'Is this student stupid or just lacking in training?' If the latter, the university will just have to work with the student. If he's trainable, just put it on that student. If he isn't, find one who is . . .

My colleague Bruce Wallace feels that Good's advice is perfectly phrased for the head instructor at the BMW or Mercedes School for Automobile Repairmen. He feels she is talking about something that is complicated (a clock, a Singer sewing machine, or a BMW automobile); students can be trained unless they are hopelessly inadequate. However, for the relatively small number of students still interested in the complex, possibly the top 1% or 2% of the student body except in the most prestigious colleges and universities, it is still possible for these students to interact with faculty primarily interested in the complex rather than the complicated. Students can be exposed to the complicated in large groups and examined in routine ways, particularly in this age of computer literacy. The complex, which requires personal interactions, can be accomplished by the senior faculty and the most dedicated students on a tutorial basis or in small discussion groups. The paradox, of course, is that most students and the families often paying their expenses want access to a burgeoning job market based primarily, but not exclusively, on the complicated. For the few students genuinely interested in mastering the complex, sufficient faculty should always be available at the personal level. This may be elitism, but, in a sense, it is not; it is satisfying the individual needs of the students for most of whom job training or qualification is the primary goal of a college or university degree.

The problem with this approach, however, whether one considers the present system emphasizing complications or the system that would enable complications to co-exist with complexity, is that most individuals will choose the complications because it offers the largest financial rewards per unit of effort. As a consequence, most of society will not be familiar with complexity but only with complications, and this is not a hospitable environment for dealing with multiple complexities simultaneously, such as interfacing product life cycles and ecological assimilative capacity. If this understanding does not come from the educational system itself at any level, how will familiarity with complexity be introduced into the larger society that must decide these issues?

## 8.6 THE ROLE OF INDUSTRY

Industry is uniquely situated to respond to the public's interest in recycling by implementing three key activities in advancing the use of recycled materials.

1. Incorporating recycled materials into more products and processes.
2. Modifying products to augment their compatibility with recycling systems.
3. Optimizing entire recycling systems in addition to optimizing individual elements or components in a system (Holl *et al.*, 1995).

There is, of course, an almost ritual complaint that it is more expensive to recycle than to use raw materials. Regrettably, this is all too frequently true because of government subsidies in the extraction or removal of the raw products, such as timber or ore, and the failure to incorporate true ecological restoration of the damaged areas into the cost of production. Another reason for not recycling is that some products contain potentially toxic materials or materials awkward to handle in the recycling process because they were not designed to be amenable to recycling when they were first manufactured. Readjusting to this new state of affairs will obviously be traumatic, but a major paradigm shift is occurring in industrial society regarding the reuse of resources, and the industries and societies that develop this industrial capability most rapidly will clearly be winners in the global marketplace as raw materials get more and more expensive for a variety of reasons. Of course, since industry is profit driven, there must be incentives and mandates.

## 8.7 CONCLUDING REMARKS

Accumulation of wastes of either technological or biological origin is deleterious to both human and ecosystem health. Furthermore, the life support system for human society has always had an ecological component and, since the agricultural and industrial revolutions, a technological component as well.

However, the concomitant explosive rate of increase in global human population size and extraction of ecological capital (old growth forests, fossil water, top soil) have created the undesirable situation of rapid accumulation of wastes of human society. This is occurring because either the natural assimilative capacity of ecosystems has been exceeded or because the waste produced by present technology, such as plastics, is not amenable to reincorporation into natural systems. Biological wastes of other species, on the other hand, have a high assimilative capacity because natural systems have become accustomed to them. The $I = P \times A \times T$ of Holdren (1986) provides some guidance for effective interfacing of human society to natural systems so that their assimilative capacity is neither impaired nor exceeded. In this equation, environmental impact equals population size × level of affluence × level of technology.

Of course, first stabilizing and then reducing the global human population to a long-term sustainable level would, over geological time, permit more humans to live on the planet than if the resources are used in a brief flash. But, human population adjustments can be more astute and orderly if the other portions of the multiplicative $I = P \times A \times T$ equation are given serious attention. For example, as is obvious in the United States, beyond some point, material possessions do not necessarily increase either literacy or the quality of life. The quality of life could arguably be improved if there were less air pollution, water pollution and more robust natural systems for recreation. Technology can either be rust belt technology or a technology that pays attention to waste minimization through efficient use of energy, reduction of process wastes and the development of life cycle products easily reintroduced into the biosphere from which they came. Of course, to the extent that wasteful practices cost an industry money, attention is paid. However, when the cost can be shifted to society (land fills, for example), it often is not a motivating factor. By incorporating the true costs of products, both the total amount of wastes and the potential toxicity can be simultaneously reduced, and the assimilative capacity of natural systems neither impaired nor exceeded. Although case histories of this sort are exceedingly rare, a model would be the enormous increase in energy efficiency in Japanese industry resulting from the first oil crisis in the early 1970s. A number of multinational corporations are also paying close attention to both waste minimization and the ability to reincorporate both production waste materials and the product itself either back into the technological cycle or into the natural cycle or a combination of the two.

The assimilative capacity for societal wastes is not infinite and is further reduced by the enormous ecological destruction that has occurred during the past century. The trends of ecological destruction and human population growth have provoked a number of warnings from mainstream science. These trends threaten both human and ecological health. An attempt to recognize that human society and natural systems are undergoing a co-evolution and that this can be either harmful or beneficial to both is now becoming clear. Paying more attention to the existence of an interface between these two complex multivariate systems is more likely to ensure a sustained beneficial interaction, rather than ignoring the interface, as has been done in the past.

*ACKNOWLEDGMENTS*

I am indebted to Bruce Wallace and B. R. Niederlehner for comments on several drafts of this manuscript. Teresa Moody transcribed the dictated first draft and Darla Donald provided much appreciated editorial assistance.

## REFERENCES

Barzun, J. (1964) *Science: the Glorious Entertainment*, Harper and Row, New York.

Cairns, J., Jr (1989) Restoring damaged ecosystems: is predisturbance condition a viable option? *Environ. Prof.*, **11,** 152–9.

Diamond, J. (1992) *The Third Chimpanzee: the Evolution and Future of the Human Animal*, Harper Collins, New York.

Diamond, J. (1994) Ecological collapses of ancient civilizations: the Golden Age that never was. *Bull. Am. Acad. Arts Sci.*, **XLVII**(5), 37–59.

Holdren, J. (1986) Too much energy, too soon. *Tech. Rev.*, January, p. 118.

Holl, K., Cairns, J. Jr and Rattray, T. T. (1995) Recycling by design, *Spec. Sci. Tech.*, **17**(2), 129–34.

Office of Technology Assessment (1993) *Harmful Non-indigenous Species in the United States*, OTA, Washington, DC.

Tunnicliffe, V. (1992) Hydrothermal vent communities of the deep sea. *Am. Sci.*, **80,** 336–50.

# PART FOUR

## *The Risk Assessment Paradigm*

# 9 *Ecological and human health risk assessment: a comparison*

JOANNA BURGER and MICHAEL GOCHFELD

## 9.1 INTRODUCTION

The public, various conservation and preservation societies, and the government increasingly are interested in the well-being of individual organisms, populations and ecosystems. This interest has manifested itself in the establishment of several government agencies devoted to the preservation, protection and management of our natural environment, federal and state regulations and laws to protect the environment, and public advocacy groups devoted to specific aspects of environmental protection.

In this chapter we examine the applicability of the traditional four-part risk assessment paradigm (hereafter called the human health risk paradigm) to ecological risk assessment. We describe the process of assessing human health risk, explore the similarities and differences between human health risk assessment and ecological risk assessment, and discuss what changes are necessary to adapt human health risk assessment criteria to suit ecosystems and their component parts. We focus attention on the hazard identification phase, and its relationship to the target identification phase of ecological risk assessment, which we approach from the viewpoint of the ecologist rather than the ecotoxicologist or environmental scientist. Although there are some differences between ecological and human health risk assessment at all steps of the process, we focus attention on the earliest steps where the differences are most profound. This is particularly important if decisions regarding applications or interpretation of ecological risk assessment are to be made by people steeped in the human health risk tradition.

Although the federal Environmental Protection Agency has often focused on non-human organisms and environmental alterations, the term 'environ-

*Interconnections Between Human and Ecosystem Health.* Edited by Richard T. Di Giulio and Emily Monosson. Published in 1996 by Chapman & Hall, London. ISBN 0 412 62400 1.

mental protection' sometimes has been used by people and agencies interested primarily in how environmental problems affect human health, and the natural environment itself has suffered in consequence. We argue that 'environmental protection' should more generally encompass not only humans, but other species, their ecosystems and the abiotic environment they all share. We think that the intimate connection between human health and the quality and stability of the surrounding environment has finally been recognized.

Approaches to environmental protection have focused mainly on species low on the food chain and on indicator species, not on community and ecosystem level approaches. When ecosystem approaches were developed, the considerations were often short-term (e.g. NEPA). Efforts to protect the natural environment led to:

1. the development of environmental impact statements under the National Environmental Policy Act (NEPA) of 1969;
2. the development of laboratory-based bioassay systems (Zeeman and Gilford, 1993);
3. the emergence of ecotoxicology which provided the backdrop for ecological risk assessment.

Government regulations led to the development of several specific bioassays for laboratory assessment of chemicals in water, which in turn played a role in the registration of pesticides and regulation of chemicals. This led to elaborate hazard assessments based heavily on laboratory tests on indicator species (National Research Council, 1991), and more recently, to the development of a context for ecological risk assessment (Norton *et al.* 1992; Suter, 1993a). In the early 1980s the importance of monitoring large geographical areas within the United States was recognized, and the National Oceanic and Atmospheric Administration created the National Status and Trends Program to monitor spatial and temporal trends of chemical contamination and biological responses to that contamination (O'Connor and Ehler, 1991). In the late 1980s, the United States Environmental Protection Agency developed the Environmental Monitoring and Assessment Program (EMAP), which not only took the indicator species concept out of the laboratory, but established a national framework for monitoring spatial and temporal trends in environmental contamination (Hunsaker *et al.* 1990a,b; Bretthauer, 1992). Although the original intent was to include higher order systems, the focus has been on individual indicator species, not communities or ecosystems.

Under NEPA environmental impact statements were required to identify the environmental consequences of proposed federal or federally funded projects. This environmental impact approach centered on the effects of proposed development projects on natural populations and communities, and this sometimes involved the projected emissions of chemicals (Suter *et al.*, 1987; Burger 1994). This approach was limited, however; for the most part environmental impact statements catalogued the species present, and seldom used biological information to evaluate the effects of the proposed chemical or activity on the habitat and community. This often resulted in short-term considerations: what risk might the proposed development or chemical exposure hold for the endangered species breeding on the site, or for sensitive plants and animals? Would important or critical habitat be lost? Other interactions that resulted from the initial exposure (on predator–prey interactions, for example) and long-term sustainability were seldom addressed. Appropriate processes and methodologies to rectify this are only now being developed. Moreover, the environmental impact process seldom involved a post-development evaluation, so that the effectiveness of the entire process was not well understood (Bartell *et al.*, 1992).

Part of the problem of trying to apply human risk assessment criteria to ecosystems lies in the development of these fields. For the most part, environmental impact assessment and ecological risk assessment developed separately (Suter *et al.*, 1987). While environmental scientists were engaged in improving the use of environmental impact statements and ecologists were improving approaches for modeling and studying ecosystem processes, health professionals were developing the process of human health risk assessment (HRA), which was often labeled 'environmental risk assessment'. The paradigm for health risk assessment was codified by the National Research Council (National Research Council, 1983), and while widely accepted, has been frequently criticized by industry, communities and risk assessors themselves. Although the utility of the process generally has been accepted, there is still considerable controversy surrounding the methodology of each step, the assumptions used, how to deal with uncertainties, and how to interpret the outcomes (National Research Council, 1993).

## 9.2 APPLYING THE HUMAN RISK ASSESSMENT PARADIGM TO ECOLOGICAL PROBLEMS

The widespread acceptance of the human risk assessment paradigm has led risk assessors (initially with little input from ecologists) to apply this paradigm to assessing the 'health' of ecosystems (National Research Council, 1986; Rapport, 1989; Barnthouse, 1992), and an analogous four-step paradigm has been proposed (National Research Council, 1993). Indeed the early stages of ecological risk assessment seemed to involve independent approaches adapted from human health risk assessment, ecotoxicology and experimental ecology.

There are major problems associated with applying the metaphor and paradigm of human health to ecosystem health (Suter, 1993b). However, while ecological risk assessment is much more complex than human health risk assessment, involving as it does interacting species, there are still important analogies between the two. Ecosystems have been likened to superorganisms with internal homeostatic mechanisms that are capable of restoring ecosystem health following perturbations (but see Suter, 1993a,b).

Engineers and actuaries recognize that risk assessment is by no means a new science for it underlies the building of bridges and the selling of insurance. Risk assessment has governed activities such as the setting of tolerance limits by the Food and Drug Administration for nearly a half century. But only in the mid-1970s do we see the emergence of a literature on risk assessment applied to environmental chemical influences on human health. A major breakthrough was the rigorous separation of risk assessment from risk management (National Research Council, 1983). Risk scientists were required to operate objectively, without regard to the cost consequences of identifying risks. The risk assessor makes no decision regarding 'acceptable risk'. It is the job of the risk managers, risk regulators, and risk perception personnel to apply this information to public policy decisions (see Slovic *et al.*, 1979; Slovic, 1987; Kasperson *et al.*, 1988). Ultimately, however, risk management considerations provide direction for risk assessment.

Human health risk assessment has four steps (see Table 9.1): hazard identification, dose–response determination, exposure assessment and risk characterization (National Research Council, 1983; EPA, 1984). To adapt human health risk assessment methodologies to ecological risk assessment, we think that the hazard identification phase requires substantial expansion, and we focus on that phase in this chapter. In our view the other steps are more analogous to the human health risk assessment process, although the additional complexities associated with ecosystems demand modification of these steps as well. For purposes of ecological risk assessment, the dose–response phase has been re-named more generally the exposure–response phase (National Research Council, 1993).

## 9.3 DIFFERENCES AND SIMILARITIES BETWEEN HUMAN HEALTH AND ECOLOGICAL RISK ASSESSMENT

Humans are, after all, a single species, albeit one which powerfully manipulates and alters its environment. The human health risk paradigm can be applied to other species, on a species-by-species basis. Ecotoxicologists have used this approach to develop a measure of ecosystem health, using indicator species (National Research Council, 1991). In most cases these indicators were advocated as 'early warning systems' to evaluate environmental quality *vis-à-vis* human health. For example, the rate of tumors in wild populations of fish may be used as indicators of the well-being of fish populations as well as to assess water quality (Baumann, 1992). Several indicator species are used by regulatory agencies, and a series of toxicity bioassays have been developed to determine the impact of particular chemicals on these species. In this case a suite of species acts as an indicator for other species (humans among them, O'Connor and Dewling, 1986).

Both human health risk and ecological risk are concerned with populations. However, for human health risk assessment it is recognized that the individual

**Table 9.1** Steps in health risk assessment* and analogies with ecological risk assessment

| Step | Human health risk assessment | Ecological risk assessment |
|---|---|---|
| Hazard identification | Identify stressor(s)<br>Identify pathophysiologic endpoints | Identify stressor(s)<br>Identify individual, community or ecosystem endpoints |
| Dose–response analysis | Review human toxicology data if available or in relevant animal species | Review toxicology data in relevant species |
| Exposure assessment | Measure contaminants in environmental media (air, soil, water, food)<br>Estimate uptake thru GI tract, skin, lungs<br>Estimate dose delivered to target organs | Measure contaminants in environmental media and biota<br>Estimate movement in food chains<br>Estimate dose to target organisms |
| Risk analysis | Use low dose extrapolation or safety factors | Estimate morbidity, mortality, recovery |
| Limitations | Toxicology data base<br>Exposure assumptions<br>Dose–response curve shape at low dose<br>Threshold assumption | Toxicology data base<br>Exposure assumptions<br>Ecosystem characterization |

* After National Research Council, 1983, 1993.

human beings are important, that no one wants to be that hypothetical one in a million individual who is the excess cancer death. In contrast, we argue that the individual organism is probably irrelevant in ecological risk assessment. It is the population that is of concern to the ecologist when assessing risks of particular stressors on ecosystems. When we as scientists or society at large undertake an ecological risk assessment for some chemical hazard or other activity such as development, we are largely concerned with the continuation of viable populations of organisms and sustainability of their ecosystem(s).

Ecological research has clearly indicated that there are species differences in patterns of life cycle, growth, survival and reproduction that have important consequences for ecological risk assessment. Microscopic invertebrates are not necessarily predictive for species high on the food chain, nor does one fish species necessarily respond exactly like another (Peakall, 1992). The controversial problem of extrapolating from rodent toxicity studies to human risk (Schneiderman *et al.*, 1975; Anon, 1993) pales by comparison with extrapolating risk estimates among species in natural ecosystems (Okkerman *et al.*, 1991). Extrapolation from experimental data is not simply a problem of interspecific comparisons. There are also problems associated with extrapolating between laboratory and field studies, and more seriously between laboratory toxicity

**Table 9.2** Similarities and differences between health and ecological risk assessment. Shown are some examples, the list for indicators is obviously not meant to be exhaustive

| | Human health risk | Ecological risk |
|---|---|---|
| Endpoint (levels) | | |
| Individual | Organ disease<br>Gross abnormalities | Organ disease<br>Gross abnormalities |
| Population | Epidemics, disease rates<br>Age structure | Epidemics, disease rates<br>Age structures<br>Habitat suitability |
| Ecosystem | | Productivity/energy flow<br>Biodiversity<br>Effects of other species and habitats on abundance<br>Trophic level composition<br>Keystone species |
| Indicators | Biomarkers<br>Physiologic changes<br>Behavioral changes<br>Reproductive outcomes | Biomarkers<br>Physiologic changes<br>Behavioral changes<br>Reproductive success |
| Boundaries | Individual<br>Determined by project or exposure or politics | Ecosystem not always discrete or closed |
| Temporal scale | 70-year 'lifetime' | Minutes to centuries |
| Spatial scale | Human body | Microscopic to global |
| Successional scale | Individual person | Dynamic nature<br>Different vulnerability of successional stages and climax |
| Biological control | Neural and endocrine systems | Species interactions |
| Goal | Maintain healthy population | Assess viability or recovery of system<br>Interactions with economics<br>Sustainability |

data and higher order ecosystem effects (such as decreased productivity or biomass, see O'Neill *et al.*, 1982).

The limitations of human health risk assessment approaches become apparent when applied to assemblages of species (communities) or to ecosystems. Some of the major problems involved in the hazard identification phase are determining boundaries, identification of target populations, the choice of endpoints, the designation of indicators, recognition of temporal and spatial scales and the impact of successional stages (Table 9.2).

For human health risk the endpoints are typically the incidence of disease or rates of mortality at the population level, while attention can be paid to diseases of specific organ systems and exposure at the individual level. Risk is expressed either as the probability of excess disease in a population under a given exposure scenario or as the level of exposure below which excess disease should not occur.

For ecological risk assessment there are population and ecosystem endpoints which are derived from ecosystems theory. These include biodiversity, species richness, abundance, keystone species, productivity, trophic level composition, energy flow, biogeochemical cycles and landscape mosaic patterns among others (Sheehan, 1984). These are attributes of ecosystems, not of individuals or single species. For example, by definition, a change in the population of a 'keystone' species influences, either positively or negatively, the populations of several other species including its prey and its competitors. To the extent they can be predicted, these features can serve as 'indicators' of ecosystem well-being, just as prevalence of pathophysiological changes or diseases are indicators of the health of human populations.

Since we cannot assess the risk of any given chemical or stressor on every organism within an ecosystem, it is usually more practical to choose indicator species and ecosystem indices for evaluating the impact of stressors. Herein we define a stressor most generally as any chemical, physical or biological agent that can induce adverse effects on ecological components (individuals, populations, communities, ecosystems, see Norton *et al.*, 1992).

## 9.4 DIFFERENCES BETWEEN HUMAN HEALTH AND ECOSYSTEM RISK ASSESSMENT IN HAZARD IDENTIFICATION

Human health risk assessment involves identifying the hazardous agents to be investigated, establishing the effects of each, and identifying the target populations. In ecological risk assessment the identification of the target population becomes elaborated as a separate phase. As we envision it, this involves at least five separate processes which are closely interrelated, each providing feedback to the others. Target identification can be influenced by hazard identification or can proceed independently of it, once a project is identified and the scope of the required risk assessment has been established.

The following sections highlight differences specifically related to spatial scales, temporal scales and stages of ecological succession which may figure prominently in identifying the target for ecological risk assessment.

### *9.4.1 SPATIAL SCALES*

The boundaries of a human are clear, but the limits of a target human population are not always apparent and arbitrary definitions or hypothetical populations are often established. Likewise, the boundaries of ecosystems are not clear, and in many cases (except the biosphere), they are arbitrary. We look at a geographical area and decide the boundaries of the target ecosystem, usually based on a habitat discontinuity. When the project is identified, the boundaries may be mandated by the need for an environmental impact statement or other legal requirement, a function of public interest (such as development of a tract of land), or may be the choice of ecologists or other scientists studying some aspect of ecosystem well-being. In most practical applications, ownership rather than ecology dictates boundaries. This choice of boundaries has clear implications for the utility of ecological risk assessments because the incorrect choice of an ecosystem boundary can have serious consequences for both the well-being of the ecosystem and for human health. For example, establishing boundaries that do not include an underlying aquifer would exclude ground water contamination from consideration, and may result in a seriously flawed assessment.

However, the issue of spatial scale for risk assessment is not the same as establishing boundaries. The spatial scale depends on the endpoints selected. Some endpoints apply over very small areas within one or more ecosystems, while others extend far beyond the direct reach of the proposed project (e.g. effects on river flow or estuary contamination).

Ecosystems can be nested: ponds within landscapes, landscapes within regions, regions within the global ecosystem. Clearly regional scale risk assessments, still very difficult to perform, are essential to meet societal goals (Hunsaker *et al.*, 1990b; Graham *et al.*, 1991), yet these often fall outside the purview of governments or agencies performing such assessments.

### *9.4.2 TEMPORAL SCALES*

The temporal scales for human risk assessment are relatively simple (Burger and Gochfeld, 1993; Gochfeld and Burger, 1993). For humans it is customary, for example, to invoke a 70 year 'lifetime' exposure (although shorter exposure scenarios are also used) and to speak in terms of lifetime risk. But for ecosystems, the lifespans of all the component organisms differ markedly, and may range from only a few minutes for some bacteria to dozens of years for sea turtles and centuries for redwoods. The time course of exposure for ecological risk assessment becomes difficult to determine when this complexity is considered.

Over the past 40 000 years the face of North America has oscillated between nearly tropical and nearly Arctic conditions, none of which resemble the snapshot of today's environment. It is obviously fruitless for risk assessors to worry about ecosystem shifts of this magnitude and time scale, yet to concern oneself with very short time scales may ignore the capacity of ecosystems to recover prior (and desirable) characteristics. This represents a major difference between the two risk assessment processes.

### *9.4.3 SUCCESSIONAL STAGE*

A specialized feature of temporal scale is the issue of successional stage which does not normally enter the human health risk process. Risk assessors seldom look beyond the lifespan of current individuals or their immediate offspring.

However, in the natural world, ecosystems are dynamic, they are constantly changing as a result of climate and other physical conditions and biological factors. In a simplistic scenario a barren space is invaded by minute pioneering species that break down rock, augment the soil and add nutrients which will subsequently support mosses and grasses. These in turn accumulate debris and set the stage for other species (old field species, shrubs, bushes) until a mature forest appears. There is a continuum of community types that might naturally exist on any geographical area, the product of the climate, soil and biology.

Under natural conditions the progression to 'climax' growth is not inexorable for there may be setbacks due to earthquakes, volcanoes, fire, or human development, re-setting the successional cycle to some earlier stage. Many otherwise healthy, natural ecosystems are maintained by fire (for example, the New Jersey Pine Barrens, see Robichaud and Buell, 1973) and change or disappear when humans control fires. On a longer time scale, climate changes cause shifts in the successional continuum in a given geographical area. Evaluating ecological risk is thus complicated because some chemicals or other anthropogenic stressor might adversely affect some, but not all successional stages. Moreover, specific chemicals or other stressors might arrest succession, or catapult the system into an earlier or later stage.

The importance of the dynamic nature of ecosystems cannot be overemphasized. Human health and well-being depend upon the 'right mix' of different successional stages. Agriculture is an early successional stage, while mature forests (that can be logged) are later successional stages. Moreover, someone has to decide what the desired successional stage or mix of stages is for any piece of land. Should this be the farmer, the farmer's neighbor, the conservationist, regulator or policy-maker? This determination is important prior to performing an ecological risk assessment.

### *9.4.4 RESILIENCY*

Although little studied, resiliency is a measure of the extent to which an ecosystem can regain its original structure and composition or some other acceptable or desirable state, after a perturbation. Natural succession following fires is an example of resiliency. At present this has little place in human risk assessment, since the targeted negative change is so unacceptable that the thought of relying on a measure of recovery seems anathema. Some human health risk assessors consider reversible, physiological changes as trivial, and not appropriate as endpoints. In this regard ecological risk assessment has something to teach health risk assessment.

## 9.5 RISK MANAGEMENT AND THE COMPARISON OF GOALS

The usual goal of health risk assessment is to evaluate the effect of a particular chemical or anthropogenic stressor on health (which, to paraphrase the World

Health Organization, can be defined as an overriding state of well-being, including the absence of disease). It should be noted in passing, however, that there is some discussion of the meaning of 'health' even for the health professional. Health or a healthy state is not simply the absence of disease but includes biological, psychological and social dimensions (Kleinbaum *et al.*, 1982). With risk information on effects of stressors on health, the public, regulators and public policy agencies can decide whether to reduce risk, by how much, and how to do so.

Ecological risk assessment is often construed to focus on ecosystem health. However, defining the well-being of ecosystems is a very challenging task. It is fraught with the following difficulties.

1. Are most existing ecosystems presently in a state of well-being? Which ones are most in need of attention?
2. Who will define which ecosystems (or components of ecosystems) should be preserved? The one that existed before the appearance of the stressor, the one that existed 100, 500, 5000 years ago, or the one that someone would like there?
3. Is change bad? And bad for whom? What components of the system are bad or good?
4. Is sustainability a viable endpoint? If so, sustainable for what? A wheat field, a managed fish pond, or a desert may be sustainable, but these may not be the desired ecosystem.
5. What index of 'well-being' can we use for ecosystems? Health has measurable objective indices such as blood pressure, heart rate, temperature, and 'quality of life,' but such measures are less obvious for ecosystem health.

One might argue that ecosystems have measures such as biodiversity, productivity and stability. But should there be a dimension of 'naturalness' as well? For example, at present New Zealand and Hawaii boast relatively high biodiversities of introduced birds and plants which, however, thrive at the expense of the unique endemic flora and fauna which are vanishing. In these cases, preservation of endemism, rather than biodiversity *per se*, are the desired goals.

It is apparent that the decision points are part of the risk management process and that these decisions provide the challenge to the risk assessor. For both health risk and ecological risk assessment, the goals are to tell the risk manager (either quantitatively or qualitatively) the conditions under which a desirable outcome can be achieved or an undesirable one prevented.

The value-laden reliance on the concept of 'sustainability' as a goal for ecological risk assessments does not eliminate the need for public policy decisions, since sustainability must be defined for specific objectives.

As mentioned above, there is feedback between the risk management and risk assessment process. The manager defines the sphere of concern (and may set a monetary limit), and the assessor identifies the relevant data and estimates the likelihood or magnitude of harm. Defining the goal for health risk assessment is thus similar to defining the goal for risk assessment of an endangered or threatened species: we want the species to survive with healthy, viable, stable populations. There will always be a need for performing these risk assessments on individual species (including humans) or on assemblages of species. However, defining the goal for ecological risk assessment for ecosystems is far more difficult.

## 9.6 PARADIGM SHIFTS FOR ECOLOGICAL RISK ASSESSMENT

The difficulties addressed above illustrate the importance of examining the basic tenets of ecological risk assessment, and evaluating the usefulness of adopting the health risk paradigm for estimating ecological risk. The four steps of health risk assessment (see Table 9.1) can provide a framework, or at least an analogy, for ecological risk assessment because they evaluate aspects of hazard identification and exposure that must be understood in the context of ecosystem structure and function, or they involve dose–response data and risk characterization that must at least be understood at the level of individual species. These steps are not defined further in this chapter (but see National Research Council, 1983, 1993). However, several additional steps and shifts in thinking are required.

### *9.6.1 REQUIRED SHIFTS IN CONCEPTUALIZATION*

The idea that protecting health will protect ecosystem health or well-being is increasingly coming under fire (Travis and Morris, 1992). The widespread rise of the term 'ecosystem health' and the application of the health risk assessment paradigm to ecological risk assessment both require shifts in conceptualization for ecological risk assessment to move forward. Suter (1993b) has criticized the term 'ecosystem health,' arguing persuasively that ecosystems are not superorganisms that have a clearly defined 'health' and that the indices normally used for assessing ecosystem health (e.g. species diversity, index of biotic integrity) are difficult to interpret. Rather he argues for measuring real properties of ecosystems (such as productivity). However, even productivity is not a simple variable. A high level of productivity can be achieved in monocultures (such as algae blooms), and pests such as water hyacinth and kudzu vines, but these are both unstable and undesirable conditions.

For ecosystem risk assessment to develop, the following shifts in thinking seem useful.

1. Ecosystems are not superorganisms, although they have homeostatic mechanisms that may return them to a 'healthy' state. Ecosystems tolerate perturbations, changes in climate for example, but there are limits or 'thresholds' beyond which recovery is no longer possible. Although the

analogy of ecosystems as superorganisms is useful in some respects, it gives a false sense of the similarities between organisms and communities.

2. The boundary definitions take on heightened importance in the case of ecosystems, and they are more diverse (see Graham *et al.*, 1991).
3. Indicator species are useful in some situations, but extrapolation between invertebrates and vertebrates or between fish and rodents, for example, is much more problematic than extrapolating between laboratory rodents and humans. Even closely related species might show dramatically different sensitivities: for example, the strong difference in sensitivity between the toxic effects of 2,3,7,8-tetrachloro-p-dibenzodioxin on the guinea pig and the rat (Poland and Knutson, 1982).
4. Measures of ecosystem effects require careful interpretation. These include productivity, energy flows and biogeochemical cycling. Other indices, that use many different measures, such as the Index of Biotic Integrity (see Suter, 1993b), are difficult to interpret. Knowing that an index value is high or low does not immediately identify the problem or how to solve it, since the component(s) affecting the index must be identified. However, the percentage loss of integrity may be a manageable endpoint. On the other hand, direct measures, such as productivity, are based on ecosystem functions and relative changes in their value can be more directly interpreted.
5. Secondary effects are important. Competitive interactions between species can be shifted by change in environmental conditions as originally shown in Birch's (1953) classic experiments on the effects of temperature on competition between two grain beetles. Similarly, factors which directly affect a keystone species can have profound effects on populations of many other species and species interactions in the ecosystem (Paine, 1966).
6. Dose–response data derived from one species may not be directly related to another species in terms of the relationship, threshold level, or in the magnitude of the response. Moreover, dose–response measures for ecosystems may not be generalizable from one ecosystem to another. In part this derives from the potential for differential dose–response relationships among the component parts, but it also is a result of the different interactions among the organisms (predation, competition, second order interactions).
7. Man's role needs to be integrated, in addition to the chemical or other stressors being considered (Slocombe, 1993). There are very few ecosystems, indeed none if one considers the General Circulation Models for the globe, that are not affected by human activities or pollutants. Ecological risk assessments should include human dimensions in all such considerations (Turner *et al.*, 1990; Slocombe, 1993).
8. Although considerations of cost incurred with health risks are not part of the risk assessment process, time constraints associated with recovery or restoration to societal objectives should be assessed. This is particularly critical because the recovery trajectory may vary dramatically among

ecosystems, and among the component parts of ecosystems. Destruction time is much shorter than restoration time (Cairns, 1993). The new and growing fields of restoration ecology (Cairns, 1991) and ecological engineering (Mitsch, 1993) are addressing these issues. Moreover, the recent emphasis on ecological research that leads to a sustainable biosphere (Risser *et al.*, 1991) involves aspects of ecological risk assessment.

9. Health risk assessment has proceeded as if it is possible to assign probabilistic values to risks for humans. Although a similar degree of reality may exist for other species, we submit that assigning probabilities to ecological risks for levels above the individual species and population levels may result in such broad error bars or confidence intervals as to be uninterpretable. This will be particularly important for ecosystem risk analysis, where deterministic or qualitative estimates may be more likely (Norton *et al.*, 1992).
10. There is a clear role for expert judgment in the case of ecological risk assessment (Grieg-Smith, 1992). Partly this derives from the relative newness of the application of the risk paradigm to ecological risk assessment, and partly it comes from the complexity of ecosystems. There are many ecologists who have studied the structure and function of particular ecosystems in depth, and their expertise will be invaluable in the formative stages of ecological risk assessment. This is particularly true for ecosystems with long-term experimental studies, such as the Hubbard Brook forest ecosystem (Likens, 1985), or for wetlands (Pascoe, 1993).

### *9.6.2 PROPOSED ADDITIONS TO THE PARADIGM*

The shifts in conceptualization needed before the four-part paradigm of risk assessment can be applied to ecological risk assessment require the addition of a number of steps, particularly to the hazard identification phase. Although this increases the complexity of the risk assessment process, it merely reflects the complexity of ecological systems. Although one could argue whether these shifts are operational or conceptual, it is clear that many of the complexities inherent in ecological risk assessment are unique, and others might educate the health risk assessor as well.

Figure 9.1 presents our approach to revising the ecological risk assessment paradigm. New steps added to the National Research Council (1993) four-part paradigm include amplification of a target identification phase reflecting the complexity of ecosystem structure and function, and the addition of an evaluation analysis phase.

As with health risk assessment, the ecological risk assessment process is usually triggered by identification of a project, a problem or a question. Thus, risk assessments can be performed in response to a specific problem (e.g. whether to clean up a PCB-contaminated landfill) or a generic problem (e.g. acceptable levels of trihalomethanes in drinking water).

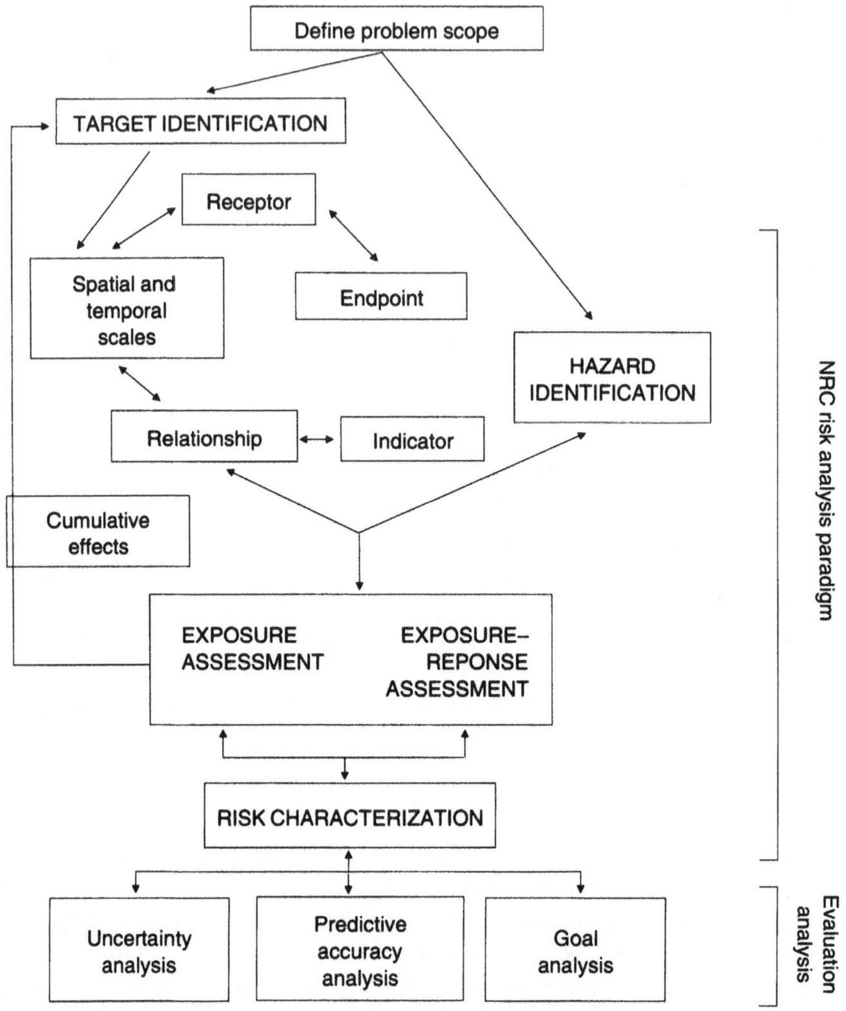


**Figure 9.1** Outline of the ecological risk assessment methodology proposed in this chapter.

## 9.7 TARGET IDENTIFICATION

In health risk assessment, hazard identification involves identification of the substance(s) of concern and their toxic effects, the target population and selection of appropriate endpoints for which risk will be estimated. For most of the 1980s this had a simple default option – cancer in humans, estimated as excess cancer deaths over a 70-year lifetime per million population (over and above a background cancer mortality rate of about 250 000 per million or 25%) (Cohrssen and Covello, 1989). Much of the risk assessment literature treats hazard identification as a simple, qualitative step, merely setting the stage, whereas in ecologic risk assessment this proves to be a highly complicated phase.

From a community and ecosystems viewpoint there is no simple default option, and the target identification of the endpoints to be examined is extremely important. Lipton *et al.* (1993) suggested that receptor identification and endpoint identification are distinct phases. Receptor identification should involve characterization of the biotic components and organization of the system; while endpoint identification should include selection of a measurable quality (population size, ecosystem function) relevant to the risk receptors, based on the responses predicted in the hazard identification phase (Keeman and Gilford, 1993; Lipton *et al.*, 1993). These steps are analogous to the identification of target populations in the hazard identification step, but their greater complexity in ecological risk assessment should not be overlooked.

Therefore, we suggest breaking up the hazard identification phase (as defined by the National Research Council, 1983) into two phases: hazard identification (the traditional identification of substance or concern with delineation of toxic effects on a range of organisms) and target identification (see Figure 9.1). The stages of target identification are partly unique to ecological risk assessment and they are intertwined.

## 9.8 ADDITIONAL STEPS IN AN ECOLOGICAL RISK ASSESSMENT

The steps outlined below are interrelated and affect one another. They do not necessarily proceed linearly, but often in tandem with feedback among the steps.

### (a) Problem statement

Risk assessments are triggered by problems that arise in society, the geographic boundaries often being dictated by ownership or jurisdiction, rather than by natural processes of the ecosystem. Usually a particular problem or project triggers a risk assessment, and the geographic boundaries for the assessment are often specified or revealed by the problem itself. Logging a forest, building an incinerator, deciding whether to remediate a hazardous waste site, all have an implicit or explicit boundary within which the risk assessor must identify specific hazards and targets. Yet in conflict situations, parties may not agree on the boundaries. For example, contamination on a military base or industrial property often results in offsite contamination, which may be of greater interest to the community or the regulators than to the responsible party.

### (b) Hazard identification

This step involves listing the hazards to be studied (e.g. incinerator emissions, ground water contamination, toxics in fish, etc.) and identifying the specific hazardous components (heavy metals, organics, pesticides, radionuclides, infectious agents). This phase includes determining the types of problems that each can cause at the biochemical, physiological, organismic, population or ecosystem level. This phase may contribute to the target identification phase, helping to identify receptors or select endpoints.

We propose that the target identification phase include the five steps described below.

1. **Spatial and temporal scale identification** This involves identification of the spatial and temporal scales of interest. The spatial scale is not necessarily congruent with the boundaries established at the outset, for the effects may spread beyond the project boundaries or may be confined to a small 'hot spot'. Spatial boundaries for ecosystem risk analyses are necessary, but risk assessors must be constantly mindful of the linkages that occur across these boundaries (Graham *et al.*, 1991).

Determining the time scale for risk assessment is difficult, and in most cases ecological risk assessments have considered the risks to organisms over only a few years, rather than the longer time scales important to population maintenance and ecosystem sustainability. Moreover, the time scale for risk assessment must take into account the normal successional stage for a given area, with the complexities inherent in changing species presence and abundance. That is, a chemical that may be used on a given site may have little effect on early successional stage species, but might affect plants and animals at later successional stages (or vice versa).

2. **Receptor identification** In this phase the risk assessor should identify the mosaic of possible biological receptors (at all levels of organization from individual to communities) that might be affected by a stressor. This requires a substantial understanding of the structure of the ecosystems.

3. **Endpoint identification** This involves determining the measures of stress to be used in the risk analysis. This selection process makes use of the information assembled from the receptor identification and sometimes from the hazard identification phase. This might include individual endpoints such as death, disease, growth, pathophysiologic changes, and biomarkers, as well as population endpoints such as rates of abnormalities, incidence of disease or death rates, or changes in reproductive success. There are also a number of community endpoints (population size, relative age classes, probability of extinction) as well as ecosystem endpoints (species diversity, productivity, biomass). The identification of surrogate or indicator species represents a combination of these two steps.

4. **Relationship identification** This involves identifying what relationships among organisms should be taken into account in risk analysis. This might include trophic level ratios, keystone predators and spatial heterogeneity analyses (Hunsaker *et al.*, 1990a), as well as secondary and higher-order

interactions among species. The functional integrity of an ecosystem and its ultimate sustainability are important values that usually must be estimated indirectly. Considerable attention has recently been devoted to landscape scale issues in ecology (Forman and Godron, 1986; Turner, 1989; Dunning *et al.*, 1991), yet these frequently have been ignored in risk assessment. Landscape scale issues include not only numbers and types of habitat patches, but size of patches, relationships among patches, and interconnectedness among patch types.

5. **Indicator identification** Where risks to individual organisms are going to be determined, either as the entire risk assessment or in combination with other community and ecosystem aspects, species must be selected for that assessment. There is a growing literature on the features that should be used to identify indicator organisms (O'Connor and Dewling, 1986; Suter, 1990; Colborn, 1991; Gilbertson *et al.*, 1991; Kremen, 1992). Species are attractive as indicator organisms if they are sensitive to environmental change (but not so sensitive that they respond to every stress), widespread (useful over a wide geographical area), can be easily and reliably examined in the field, can be assessed in a repeatable fashion, and are cost effective. Furthermore, for a species to be widely accepted as an indicator species, it must be of public concern (such as most birds and marine mammals). The five aspects of target identification (see Figure 9.1) obviously are interrelated, and with the completion of each step, the previous steps should be reviewed. Moreover, after the target identification, hazard identification, exposure assessment and response assessment phases have been performed, there should be a cumulative effects identification phase to determine that there are not any obvious cumulative effects that have been missed in the target identification phase. Cumulative effects result from the additive effects of individual chemicals or stressors applied at different times, or the interactive effect of multiple chemicals or stressors (Slocombe, 1993; Spaling and Smit, 1993). Adequate consideration of cumulative effects must take into account the long history most ecosystems have of induced changes from the most primitive hunter-gatherers who distributed seeds as a byproduct of gathering to modern-day agriculture, industry and urbanization. This is, undoubtedly, a vast project, but ecological risk assessors should at least consider the effects of past endeavors on ecosystems.

Cumulative effects identification may involve qualitative rather than quantitative risk assessments at the present time, particularly for the additive effects of different chemicals or stressors. Although probabilistic risk assessments have been performed for a number of chemicals (particularly for health risk assessments) separately, there are few such risk assessments for mixtures. Yet cumulative effects may be the most important ones for ecological systems. The fact that we have few methodologies to assess the risks from mixed exposures to ecosystems should not dissuade us from considering these risks.

### (c) Evaluation analysis

Finally there are evaluation analyses that should be added to ecological risk assessment. These include uncertainty analysis, predictive accuracy analysis and goal analysis. Uncertainty analysis (analyzing the uncertainties involved with each phase of risk assessment) is an important part of risk assessment (O'Neill *et al.*, 1982; Fogarty *et al.*, 1992). Uncertainties often include those areas where there are no data or where interspecific extrapolations must be made. There are many uncertainties associated with data-rich aspects of risk assessment. For example, although there is often a wealth of data on fish populations, there are many inherent difficulties associated with assessing the size of any population, especially one that lives out of sight below the water surface (Fogarty *et al.*, 1992). This phase should also include clear examination of the uncertainties associated with the assumptions made to compensate for inadequate data (Lucier, 1993).

Burger (1994) suggested that one important step in environmental impact assessment and ecological risk assessment is predictive accuracy: the evaluation of the relative success of the initial risk assessment. This phase should address the question: how well did the impact statement or risk assessment predict the result?

In health risk assessment where the outcome is lifetime cancer risk, evaluating the validity of a prior risk estimate is not feasible; the time course between prediction and outcome is too long. However, many of the endpoints in ecological risk assessment involve short-term changes (in species composition, relative abundance, age structure, behavioral shifts), measurable over a period of one or a few years, which can be evaluated and can provide feedback to the process. Thus the risk assessment process should define those measurements which must be made to evaluate it.

Although many risk assessments are performed to predict future effects of a particular chemical or stressor, other risk assessments are aimed at retrospectively examining the effects of some chemical already present in the environment (Suter, 1993a). In this latter type of risk assessment, predictive accuracy could be employed at the time of the ecological risk assessment.

## 9.9 SOCIETAL GOALS

Finally there should be an assessment of how well the initial societal goal for the risk assessment was met. If the initial societal objective was to preserve the integrity or sustainability of a certain ecosystem, or maintain high native species diversity, then the evaluation phase of risk assessment should address whether the methodology was sufficient to determine whether the goal was met, and ultimately society should reflect on whether the goal was achieved.

## 9.10 CONCLUSIONS

Ecological risk assessment is a relatively new discipline, but one that is critical for the preservation, maintenance and sustainability of ecosystems from the local to the global scale. Recently a variety of environmental, ecological, mathematical, and geophysical scientists have examined the applicability of the health risk assessment paradigm to ecological problems. Several authors have pointed out the difficulties of treating ecosystems as superorganisms, and of extrapolating risk assessment methodologies developed for a single species to the complexity of organisms, structures and functions that comprise an ecosystem.

In this chapter we proposed shifts in the hazard identification phase that seem prudent for ecological risk assessment, including the addition of a target identification phase that includes receptor, endpoint, relationship, indicator and spatial and temporal scale identifications. These steps will amplify the aspects of ecosystems that differ in the risk assessment process.

We also add an evaluation analysis phase that includes uncertainty analysis, predictive accuracy analysis and goal analysis. We believe that these will contribute to understanding the relative value of the ecological risk assessments.

### *ACKNOWLEDGMENTS*

We have benefited greatly from discussions with Keith Cooper, Bernard Goldstein, Susan Norton, Daniel Wartenberg, Michael Gallo and an anonymous reviewer, and we appreciate their critical reading and reflection of earlier drafts of this chapter. The authors were partly funded by NIEHS grants ES 05022 and ES 05955, by the Department of Energy under CRESP (the Consortium for Risk Evaluation with Stakeholder Participation), and by the Environmental and Occupational Health Sciences Institute.

## REFERENCES

Anon. (1993) Toxicity tests in animals: extrapolating to risk. *Environ. Health Perspect.*, **101,** 396–401.

Barnthouse, L. W. (1992) The role of models in ecological risk assessment: a 1990s perspective. *Environ. Toxicol. Chem.*, **11,** 1751–60.

Bartell, S. M., Gardner, R. H. and O'Neill, R. V. (1992) *Ecological Risk Estimation*, Lewis Press, Boca Raton, FL.

Baumann, P. C. (1992) The use of tumors in wild populations of fish to assess ecosystem health. *J. Aquatic Ecosystem Health*, **1,** 135–46.

Birch, L. C. (1953) Experimental background to the study of distribution and abundance of insects. III. The relations between innate capacity for increases and survival of different species of beetles living together on the same food. *Evolution*, **7**, 136–44.

Bretthauer, E. W. (1992) The challenge of ecological risk assessment. *Environ. Toxicol. Chem.*, **11**, 1661–2.

Burger, J. (1994) How should success be measured in ecological risk assessment? The importance of 'predictive accuracy'. *J. Environ. Health Toxicol.*, **42**, 367–76.

Burger, J. and Gochfeld, M. (1993) Temporal scales in ecological risk assessment. *Arch. Environ. Contam. Toxicol.*, **23**, 484–8.

Cairns, J. (1991) Restoration ecology: a major opportunity for ecotoxicologists. *Environ. Toxicol. Chem.*, **10**, 429–32.

Cairns, J. (1993) The balance of ecological destruction and repair. *Environ. Health Perspect.*, **101**, 206.

Cohrssen, J. J. and Covello, V. T. (1989) *Risk Analysis: a Guide to Principles and Methods for Analyzing Health and Environmental Risks*. Council on Environmental Quality, Washington, DC. (1989)

Colborn, T. (1991) Epidemiology of Great Lakes bald eagles. *J. Toxicol. Environ. Health*, **33**, 395–453.

Dunning, J. B., Danielson, B. J. and Pulliam, H. R. (1991) Ecological processes that affect populations in complex landscapes. *Oikos*, **65**, 169–75.

EPA (Environmental Protection Agency) (1984) *Risk Assessment and Management: Framework for Decision Making*, US Environmental Protection Agency, Washington, DC.

Fogarty, M. J., Rosenberg, A. A. and Sissenwine, M. P. (1992) Fisheries risk assessment: sources of uncertainty. *Environ. Sci. Technol.*, **26**, 440–7.

Forman, R. T. T. and Godron, M. (1986) *Landscape Ecology*, John Wiley and Sons, New York.

Gilbertson, M., Kubiak, T., Ludwig, J. and Fox, G. (1991) Great Lakes embryo mortality, edema, and deformation syndrome (GLEMEDS) in colonial fish-eating birds. *J. Toxicol. Environ. Health*, **33**, 455–520.

Gochfeld, M. and Burger, J. (1993) Evolutionary consequences for ecological risk assessment and management. *Environ. Monit. Assess.*, **28**, 161–8.

Graham, R. L., Hunsaker, C. T., O'Neill, R. V. and Jackson, B. L. (1991) Ecological risk assessment at the regional scale. *Ecol. Applic.*, **1**, 196–206.

Grieg-Smith, P. W. (1992) A European perspective on ecological risk assessment, illustrated by pesticide registration procedures in the United Kingdom. *Environ. Toxicol. Chem.*, **11**, 1673–89.

Hunsaker, C., Carpenter, D. and Messer, J. (1990a) Ecological indicators for regional monitoring. *Bull. Ecol. Soc. Am.*, **71**, 165–72.

Hunsaker, C. T., Graham, R. L., Suter, G. II *et al.* (1990b) Assessing ecological risk on a regional scale. *Environ. Manage.*, **14**, 325–32.

Kasperson, R. E., Renn, O., Slovic, P. *et al.* (1988) The social amplification of risk: a conceptual framework. *Risk. Anal.*, **8**, 177–87.

Keeman, M. and Gilford, J. (1993) Ecological hazard evaluation and risk assessment under EPA's Toxic Substances Control Act (TSCA): an introduction, in *Environmental Toxicology and Risk Assessment* (eds W. G. Landis, J. S. Hughes and M. A. Lewis), American Society for Testing and Materials, Philadelphia.

Kleinbaum, D. G., Kupper, L. L. and Morgenstern, H. (1982) *Epidemiologic Research*, Van Nostrand Reinhold, New York.

Kremen, C. (1992) Assessing the indicator properties of species assemblages for natural areas monitoring. *Ecol. Applic.*, **2**, 203–17.

Likens, G. E. (1985) An experimental approach for the study of ecosystems. *J. Ecol.*, **73,** 381–96.
Lipton, J., Galbraith, H., Burger, J. and Wartenberg, D. (1993) A paradigm for ecological risk assessment. *Environ. Manage.*, **17,** 1–5.
Lucier, G. W. (1993) Risk assessment: good science for good decisions. *Environ. Health Perspect.*, **101,** 366.
Mitsch, W. J. (1993) Ecological engineering. *Environ. Sci. Technol.*, **27,** 438–45.
National Research Council (1983) *Risk Assessment in the Federal Government: Managing the Process*, National Academy Press, Washington, DC.
National Research Council (1986) *Ecological Knowledge and Environmental Problem Solving*, National Academy Press, Washington DC.
National Research Council (1991) *Animals as Sentinels of Environmental Health Hazards*, National Academy Press, Washington, DC.
National Research Council (1993) *Issues in Risk Assessment*, National Academy Press, Washington, DC.
Norton, S. B., Rodier, D. R., Gentile, J. H. *et al.* (1992) A framework for ecological risk assessment at the EPA. *Environ. Toxicol. Chem.*, **11,** 1663–72.
O'Connor, J. S. and Dewling, R. T. (1986) Indices of marine degradation: their utility. *Environ. Manage.*, **10,** 335–43.
O'Connor, T. P. and Ehler, C. N. (1991) Results from the NOAA National Status and Trends program on distribution of chemical contamination in the coastal and estuarine United States. *Environ. Monit. Assess.*, **17,** 33–49.
O'Neill, R. V., Gardner, R. H., Barnthouse, L. W. *et al.* (1982) Ecosystem risk analysis: a new methodology. *Environ. Toxicol. Chem.*, **1,** 167–77.
Okkerman, P. C., Plassche, E. J. V. D., Slooff, W. *et al.* (1991) Ecotoxicological effects assessment: a comparison of several extrapolation procedures. *Ecotoxicol. Environ. Saf.*, **21,** 182–93.
Paine, R. T. (1966) Food web complexity and species diversity. *Am. Natur.*, **100,** 65–75.
Pascoe, G. A. (1993) Wetland risk assessment. *Environ. Toxicol. Chem.*, **12,** 2293–307.
Peakall, D. (1992) *Animal Biomarkers as Pollution Indicators*, Chapman and Hall, London.
Poland, A. and Knutson, J. C. (1982) 2,3,7,8-tetrachloro-p-dibenzo dioxin and related halogenated hydrocarbons: examinations of the mechanisms of toxicity. *Ann. Pharmacol. Toxicol.*, **22,** 517–54.
Rapport, D. J. (1989) What constitutes ecosystem health? *Perspec. Biol. Med.*, **33,** 120–32.
Risser, P. G., Lubchenco, J. and Levin, S. A. (1991) Biological research priorities – a sustainable biosphere. *BioScience*, **41,** 625–7.
Robichaud, B. and Buell, M. F. (1973) *Vegetation of New Jersey*, Rutgers University Press, New Brunswick, NJ.
Schneiderman, M. A., Mantel, N. and Brown, C. C. (1975) From mouse to man – or how to get from the laboratory to Park Avenue and 59th Street. *Ann. NY. Acad. Sci.*, **246,** 237–48.
Sheehan, P. J. (1984) Effects on community and ecosystem structure and dynamics, in *Effects of Pollutants at the Ecosystem Level* (eds P. J. Sheehan, D. R. Miller, G. C. Butler and P. Bourdeau), John Wiley and Sons, Chichester.
Slocombe, D. D. (1993) Environmental planning, ecosystem science, and ecosystem approaches for integrating environment and development. *Environ. Manage.*, **17,** 289–303.

Slovic, P. (1987) Perception of risk. *Science*, **236,** 280–5.

Slovic, P., Fischoff, B. and Lichtenstein, S. (1979) Rating the risks. *Environ.*, **21,** 14–20.

Spaling, H., and Smit, B. (1993) Cumulative environmental change: conceptual frameworks, evaluation approaches, and institutional perspectives. *Environ. Manage.*, **17,** 587–600.

Suter, G. W., II (1990) Endpoints for regional ecological risk assessment. *Environ. Manage.*, **14,** 9–23.

Suter, G. W., II (1993a) *Ecological Risk Assessment*, Lewis Publishing, Boca Raton, FL.

Suter, G. W., II. (1993b) A critique of ecosystem health concepts and indexes. *Environ. Toxicol. Chem.*, **12,** 1533–9.

Suter, G. W., II, Barnthouse, L. W. and O'Neill, R. V. (1987) Treatment of risk in environmental impact assessment. *Environ. Manage.*, **11,** 295–303.

Travis, C. C. and Morris, J. M. (1992) The emergence of ecological risk assessment. *Risk Analysis*, **12,** 167–8.

Turner, B. L., Kasperson, R. E., Meyer, W. B. *et al.* (1990) Two types of global environmental change. *Global Environ. Change*, **1,** 14–22.

Turner, M. G. (1989) Landscape ecology: the effect of pattern on process. *Ann. Rev. Ecol.*, **20,** 171–97.

Zeeman, M. and Gilford, J. (1993) Ecological hazard evaluation and risk assessment under EPA's Toxic Substances Control Act (TSCA): an introduction, in *Environmental Toxicology and Risk Assessment* (eds W. G. Landis, J. S. Hughes and M. A. Lewis), American Society for Testing and Materials, Philadelphia.

# 10 *Toxicological and biostatistical foundations for the derivation of a generic interspecies uncertainty factor for application in non-carcinogen risk assessment*

EDWARD J. CALABRESE and LINDA A. BALDWIN

## 10.1 INTRODUCTION

The use of an uncertainty factor (UF) to account for interspecies variation in risk assessment procedures for non-carcinogens is well known and implemented by regulatory agencies at the federal and state levels. The approach that has been widely adopted is to assume that humans may be 10-fold more sensitive than the animal model. This factor of 10 has become routinely adopted in essentially all risk assessment procedures involving animal model data for extrapolation.

Despite the long-standing use of the interspecies UF of 10, only limited biological and/or toxicological justification for the interspecies UF has ever been put forth by any regulatory agency (Dourson and Stara, 1983) or national advisory committee (e.g. National Academy of Sciences Safe Drinking Water Committee). The adoption of the 10-fold factor appears to have been based on a combination of public health protection philosophy, practical/intuitive toxicological insights based on experience, and a sense that it achieves its goal of protecting human health. The present chapter offers what we believe to be a toxicological and statistically defensible foundation for deriving the interspecies UF, its data base requirements, and statistical procedures for its derivation. In brief, the recommended interspecies UF is defined as the 95% of the population of 95% prediction intervals (PI) for binary interspecies

*Interconnections Between Human and Ecosystem Health.* Edited by Richard T. Di Giulio and Emily Monosson. Published in 1996 by Chapman & Hall, London. ISBN 0 412 62400 1.

comparisons based on phylogenetic relatedness. More specifically, the UF is derived by determining the minimum ratio of the estimated toxicity value and its 95% upper or lower PI after back-transformation from the logarithmic expression.

This chapter presents the toxicological and statistical basis for this proposal and its implications for judging the reliability of current regulatory interspecies UF procedures as well as offering a fundamentally novel approach to deriving an interspecies UF.

## 10.2 PROPOSED METHODOLOGY

An extensive data base on interspecies variation in susceptibility to toxic agents exists in the aquatic toxicology area. The toxicity data are principally, though not exclusively, based on acutely toxic responses. The data are arranged in the form of binary interspecies comparisons with respect to toxicity from dozens to over 500 agents depending on the specific binary comparison. A binary comparison in the present context involves comparing the responses of two species to agents that were tested in both species. For example, two species of fish (e.g. smallmouth bass and perch) have been used to test over 500 of the same toxicants (Figure 10.1). A binary comparison of these two species would include more than 500 agents. These data have been organized to assess whether a mathematical relationship exists such that the $LC_{50}$ of one species may be a useful predictor of the $LC_{50}$ in the other species via the use of regression modeling.

The above binary comparison methodology has been used by various authors (Suter *et al.*, 1983; Slooff *et al.*, 1986; Barnthouse *et al.*, 1990) to estimate the $LC_{50}$ for any new chemical in an untested species (e.g. smallmouth bass) if the $LC_{50}$ were known for the perch. The estimate is made by calculating a prediction interval (PI) for the unknown chemical. Barnthouse *et al.* (1990) have provided 95% PI estimates for numerous binary interspecies comparisons and organized them via phylogenetic relatedness. For example, interspecies comparisons were provided when the comparisons represented species-within-genus, genera-within-family, families-within-order, and orders-within-class comparisons. For example, in Figure 10.2 a species-within-genus comparison would represent a binary comparison of species 1 with species 2. A genera-within-family binary comparison would be represented by a comparison of species 1 with species 3. The reason for organizing the comparisons in this phylogenetic manner is the assumption that interspecies variation in susceptibility would increase as the phylogenetic distance increased.

Table 10.1 provides a summary of the data base of phylogenetically based interspecies binary comparisons. The 95% PI for each binary comparison is provided, along with the number of different chemical agents tested for each binary comparison. The weighted mean value indicates that in general the closer the animal species were related, the smaller the 95% PI. The range of

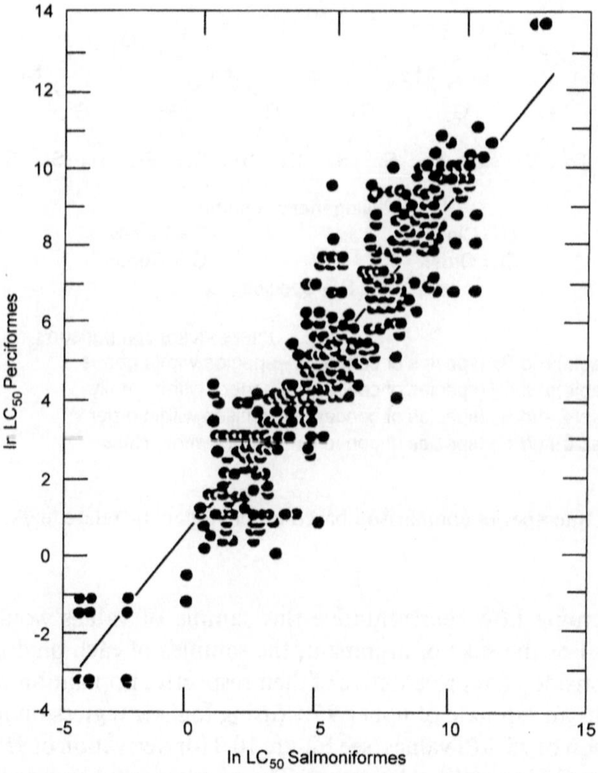


**Figure 10.1** Natural logarithms of $LC_{50}$ values for Perciformes plotted against Salmoniformes (orders of the same class, Osteichthyes). The solid line represents the least-square linear regression of the natural logarithm of $LC_{50}$ values for Perciformes species on the natural logarithm of $LC_{50}$ values for Salmoniformes species. Each circle represents the $LC_{50}$ value of a specific chemical for both species. The number of chemicals represented in the figure is 503. (Data from Johnson and Finley, 1980.)

weighted means of 95% PI is from a low of 6.0 (species-within-genus) to a high of 26.0 for the orders-within-class grouping.

Slooff *et al.* (1986) transformed the concept of the 95% PI into a 95% UF. Figure 10.3 presents a graphic foundation of the PI as well as statistical definition and relationship to the UF concept. Thus, the species-within-genus 95% UF, as anticipated, is considerably smaller than the 95% UF for orders-within-class. The magnitude of interspecies variation in 95% PI values follows fairly closely with phylogenetic relatedness, as expected. Inconsistencies such as the similar estimates for species-within-genus and genera-within-family are likely related to issues concerning representativeness, number of binary comparisons and number and nature of chemical agents tested.

The binary comparison values do not represent the population (or universe) of such values but must be considered a sample of the population. No knowledge

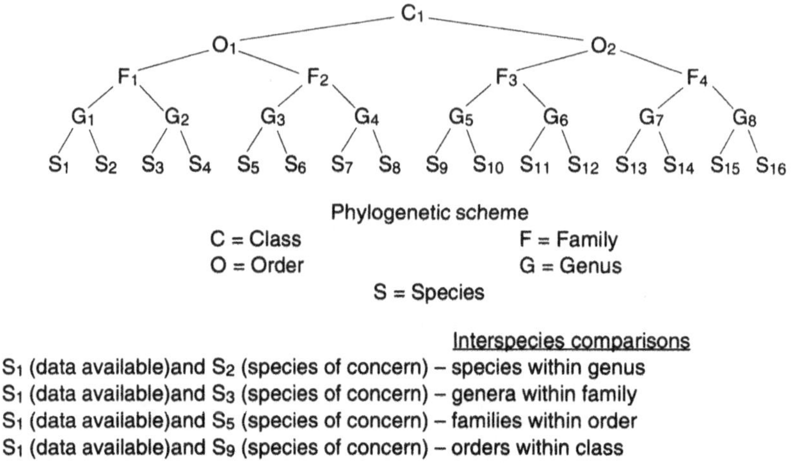


**Figure 10.2** Interspecies comparison based on phylogenetic relatedness.

exists concerning how representative this sample of values would be of the population. For the sake of argument, the samples of each phylogenetic subgroup are considered representative of their respective population values. Table 10.2 provides an estimate of upper 95% (using logistic regression modeling) of the population of 95% PI values (see Figure 10.3 for derivation of 95% PI values) according to phylogenetic relatedness. The unexpectedly high value from the families-within-order extrapolation group is partially inconsistent with the proposed phylogenetic relationship. This inconsistency is principally a result of the low number of binary comparisons ($n$ = 7) and high variability of individual estimates in the families-within-order comparison group. This value is less stable than the orders-within-class grouping. Given the amount of data, the orders-within-class comparison offers the most stable and reliable perspective. We propose that these values can be used to provide a toxicologically and statistically based foundation for generic interspecies UFs when normalized for phylogenetic relatedness. The data suggest that four different UFs be adopted according to phylogenetic relatedness. The choice of 95% UFs would range from a low of 10 for the species-within-genus to a high of 65 for the orders-within-class. The genera-within-family and families-within-order groupings are more difficult to determine. Based on the phylogenetic relatedness concept, these two groups are estimated to be intermediate between the boundary values (i.e. species-within-genus, orders-within-class), approximating 25 and 50.

The proposed methodology approach takes into account two critical components in any interspecies UF estimation process: first, the need to address the universe of species (as is done via the use of logistic regression) and secondly, the need to incorporate the new chemicals (as is accomplished via

**Table 10.1** Taxonomic extrapolation: means and weighted means calculated for the 95% and 99% prediction intervals (PI) for uncertainty factors calculated from regression models

| X variable | Y variable | *n* | Uncertainty factor 95% PI | Uncertainty factor 99% PI |
|---|---|---|---|---|
| **Taxonomic extrapolation: species-within-genera** | | | | |
| *Salmo clarkii* | *S. gairdneri* | 18 | 9 | 13 |
| *Salmo clarkii* | *S. salar* | 6 | 6 | 10 |
| *Salmo clarkii* | *S. trutta* | 8 | 6 | 8 |
| *Salmo gairdneri* | *S. salar* | 10 | 7 | 11 |
| *Salmo gairdneri* | *S. trutta* | 15 | 4 | 5 |
| *Salmo salar* | *S. trutta* | 7 | 5 | 8 |
| *Ictalurus melas* | *I. punctatus* | 12 | 5 | 7 |
| *Lepomis cyanellus* | *L. macrochirus* | 14 | 6 | 9 |
| *Fundulus heteroclitus* | *F. majalis* | 12 | 6 | 8 |
| Mean | | | 6.1 | 10.1 |
| Weighted Mean | | | 6.0 | 7.4 |
| **Taxonomic extrapolation: genera-within-families** | | | | |
| *Oncorynchus* | *Salmo* | 56 | 5 | 6 |
| *Oncorynchus* | *Salvelinus* | 13 | 4 | 5 |
| *Salmo* | *Salvelinus* | 56 | 5 | 7 |
| *Carassius* | *Cyprinus* | 8 | 4 | 6 |
| *Carassius* | *Pimephales* | 19 | 7 | 9 |
| *Cyprinus* | *Pimephales* | 10 | 7 | 10 |
| *Lepomis* | *Micropterus* | 30 | 8 | 11 |
| *Lepomis* | *Pomoxis* | 8 | 9 | 13 |
| *Cyprinodon* | *Fundulus* | 12 | 6 | 8 |
| Mean | | | 6.1 | 8.3 |
| Weighted Mean | | | 5.8 | 7.7 |
| **Taxonomic extrapolation: families-within-orders** | | | | |
| Centrarchidae | Percidae | 47 | 10 | 14 |
| Centrarchidae | Cichlidae | 6 | 4 | 6 |
| Percidae | Cichlidae | 5 | 13 | 24 |
| Salmonidae | Esocidae | 11 | 9 | 13 |
| Atherinidae | Cyprinodontidae | 32 | 7 | 9 |
| Mugilidae | Labridae | 12 | 55 | 78 |
| Cyprinodontidae | Poecillidae | 12 | 3 | 5 |
| Mean | | | 14.4 | 21.3 |
| Weighted Mean | | | 12.6 | 17.9 |
| **Taxonomic extrapolation: orders-within-classes** | | | | |
| Salmoniformes | Cypriniformes | 225 | 20 | 27 |
| Salmoniformes | Siluriformes | 203 | 39 | 51 |
| Salmoniformes | Perciformes | 443 | 12 | 16 |
| Cypriniformes | Siluriformes | 111 | 11 | 15 |
| Cypriniformes | Perciformes | 219 | 32 | 43 |
| Siluriformes | Perciformes | 190 | 63 | 83 |
| Anguiliformes | Tetraodontiformes | 12 | 13 | 18 |
| Anguiliformes | Perciformes | 34 | 25 | 34 |

**Table 10.1** Continued

| | X variable | Y variable | *n* | Uncertainty factor 95% PI | 99% PI |
|---|---|---|---|---|---|
| | Anguiliformes | Gasterosteiformes | 8 | 16 | 24 |
| | Anguiliformes | Atheriniformes | 46 | 9 | 12 |
| | Atheriniformes | Cypriniformes | 7 | 501 | 786* |
| | Atheriniformes | Tetraodontiformes | 46 | 13 | 17 |
| | Atheriniformes | Perciformes | 148 | 25 | 33 |
| | Atheriniformes | Gasterosteiformes | 36 | 20 | 27 |
| | Gasterosteiformes | Tetraodontiformes | 8 | 20 | 30 |
| | Gasterosteiformes | Perciformes | 33 | 32 | 43 |
| | Perciformes | Tetraodontiformes | 34 | 25 | 34 |
| Mean | | | 23.5 | 31.7 | |
| Weighted Mean | | | 26.0 | 34.5 | |

* Not included in calculations.

the use of the PI approach). These findings and interpretations are based directly on data derived from acute toxicology experiments in fish. It assumes that the concept of phylogenetic relatedness in relationship to toxicity that is seen within fish species would apply to mammals and that the magnitude of the phylogenetic differences observed among fish species would be quantitatively comparable to mammalian toxicology.

## 10.3 DISCUSSION

The proposed methodology offers a number of important strengths in providing a foundation for the interspecies UF derivation:

### (a) Strengths

1. It represents an extensive data base obtained via a standardized testing protocol with respect to a critical integrative endpoint (i.e. $LC_{50}$).

**Table 10.2** Upper 95% uncertainty factors calculated for the 95% and 99% prediction intervals (Table 10.1)

| Regression model | Prediction interval 95% | 99% |
|---|---|---|
| Species-within-genera extrapolation | 10.0 | 16.3 |
| Genera-within-families extrapolation | 11.7 | 16.9 |
| Families-within-orders extrapolation | 99.5 | 145.0 |
| Orders-within-classes extrapolation | 64.8 | 87.5 |

Based on the scheme of Van Straalen and Denneman (1989).

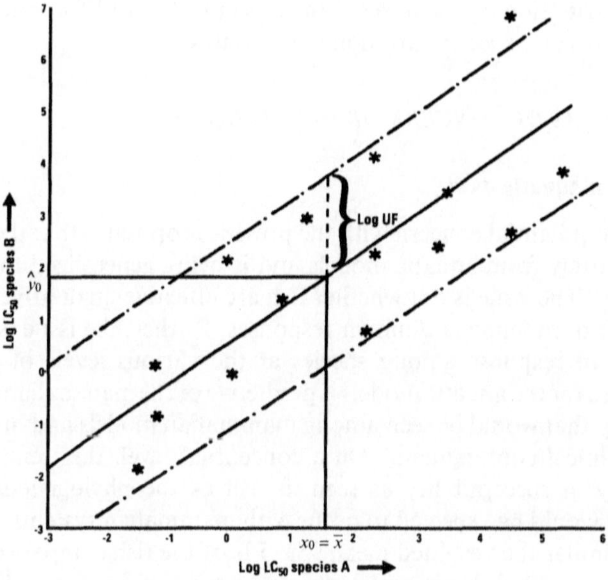


$$\hat{y}_0 \pm t \cdot s \left(1 + \frac{1}{n} + \frac{(x_0 - \bar{x})^2}{\sum_{1}^{n} (x_i - \bar{x})^2}\right)^{1/2}$$

where $t$ = 1/2 percentile of a Student distribution with $n - 2$ degrees of freedom; $s$ = estimated residual variance; $n$ = number of observations; $x$ = known log $LC_{50}$ species A; and $\hat{y}$ = estimated log $LC_{50}$ species B. When $x_0 = \bar{x}$, the prediction interval becomes $\hat{y}_0 \pm t \cdot s[1 + (1/n)]^{1/2}$. The uncertainty factor is defined as the minimum ratio of the estimated toxicity value and its 95% upper or lower prediction limit after back transformation: $UF = 10^{t \cdot s}[1 + (1 + (1/n)]^{1/2}$. In applied terms: If the toxicity of a given compound for A is known, the value for B is in the range of A/UF < B < A · UF with a probability of 95%.

**Figure 10.3** Determination of uncertainty factors (UF). (Reproduced with permission from Slooff *et al.*, 1986.)

2. It offers the capability to incorporate phylogenetic relatedness to the predictive endpoint, which represents a significant advance and is entirely consistent with the biologically persuasive evolutionary paradigm of modern molecular biology relating genetic factors to susceptibility and/or resistance to chemical insults.
3. The data base considers a large number of species representing different sizes, various biological adaptations, and variation in susceptibilities.
4. The data base is composed of assessments of more than 400 different chemical agents representing several dozen chemical classes (e.g. pesticides, metals, PAHs, etc.). The data base has the capacity to provide strong generalizations to account for both inherent species variation and large numbers of chemical agents.
5. The data base permits the application of statistical evaluation to describe

the distribution of responses with respect to both PI for specific chemical responses and species variation in responses.

### *10.3.1 AREAS OF CONCERN OR DISCUSSION*

#### (a) Use of aquatic data

An area of potential concern with the present proposal is that the data base is drawn entirely from aquatic models and is being generalized to mammalian phylogeny. The issue is not whether fish are effective qualitative/quantitative predictors of mammalian/human responses. Rather, the issue is whether the variation in response among species at the various levels of phylogenetic relatedness for the aquatic models is predictive of the mammalian phylogenetic variability that would be seen among mammalian models and humans for the same chemical contaminants. On a conceptual level, the trend in increased variability in susceptibility as seen in fish as the phylogenetic relatedness decreases would be expected to occur with mammalian systems. How quantitatively similar the weighted mean 95% PIs of the fish comparisons would be for the various phylogenetic relatedness comparisons in mammals is unknown. However, the use of biological systematics to provide a common measure of evolutionary/biological relatedness among the various animal classifications (e.g. fish and mammals) is a valuable and powerful tool that has rarely been applied to the field of toxicology/risk assessment. For example, the basic unit of comparison, the species, is similarly defined in fish as well as mammals. Although less precise than the species concept, the same conceptual definition proceeds to broader categories (genus to class) across the animal kingdom. Thus, the trend of interspecies variability observed in various phylogenetic related categories in fish would be expected to be qualitatively similar in mammals as well.

#### (b) Use of acute toxicity

Another area of possible concern is that the data base uses acute rather than chronic toxic responses. This does not appear to be a serious concern because acute responses have been shown to be effective predictors of chronic effects of both a carcinogenic (Zeise *et al.*, 1986) and non-carcinogenic nature (Kenaga, 1982; Sloof *et al.*, 1986; Layton *et al.*, 1987; Barnthouse *et al.*, 1990). In fact, the chronic no observed adverse effect level (NOAEL) in mammalian models and the chronic maximum acceptable toxicant concentration in fish have been similarly estimated by dividing the acutely lethal dose ($LD_{50}/LC_{50}$) by approximately 50–75 (Kenaga, 1982; Sloof *et al.*, 1986; Layton *et al.*, 1987; Calabrese and Baldwin, 1993). These data show a high degree of fundamental concordance between fish and mammalian responses with respect to the capacity of acute doses to estimate chronic responses.

### (c) What does the 95% UF number represent?

There is a need to define the biological and statistical meaning of the interspecies UF. The 95% UF as described here represents the upper 95% of the distribution of binary interspecies comparison 95% PI values. This is interpreted as 95% of experiments in which a chemical is tested would respond within the given PI (i.e. 95% PI). This also is interpreted to mean that 95% of every 100 unknown chemicals tested would display a response within the calculated PI. The 95% PI can, therefore, be used as a measure of interspecies variation. We then estimate the upper 95% of these 'individual measures of interspecies variation' (i.e. the distribution of the 95% PI). This is then collectively interpreted as the following: 95% of chemicals would not exceed a given PI in 95% of species tested. The risk assessor has the flexibility to change the size of the PI as well as that portion of the logistic distribution deemed suitable for UF selection. For example, if the 99th percentile of the population of 95% PI were selected for the UF, then the range of phylogenetic UFs would be increased from the 10- to 65-fold range to the 16- to 87-fold range (see Table 10.2). The final selection of which range of UF values to select would be based on value judgments.

### (d) Size of rodent to human UFs via proposed scheme

The field of mammalian toxicology in which mice, rats, gerbils, guinea pigs, cats and dogs are used as models to estimate human responses represents orders-within-class comparisons. Using the scheme outlined above suggests that the UF for such comparisons could range from 65 to 87 (possibly rounded to 50–100) rather than the 10-fold value currently used, depending on which quantitative estimate for UF derivation were selected.

## REFERENCES

Barnthouse, L. W., Suter, G. W. and Rosen, A. E. (1990) Risks of toxic contaminants to exploited fish populations: influence of life history, data uncertainty and exploitation intensity. *Environ. Toxicol. Chem.*, **9**, 297–311.

Calabrese, E. J. and Baldwin, L. (1993) *Performing Ecological Risk Assessments*, Lewis Publishers, Chelsea, MI.

Dourson, M. L. and Stara, J. F. (1983) Regulatory history and experimental support of uncertainty (safety) factors. *Reg. Toxicol. Pharmacol.*, **3**, 224–38.

Johnson, W. W. and Finley, M. T. (1980) *Handbook of Acute Toxicity of Chemicals to Fish and Aquatic Invertebrates.* US Fish and Wildlife Service Resource Publication 137. US Department of the Interior, Washington, DC.

Kenaga, E. E. (1982) Predictability of chronic toxicity from acute toxicity of chemicals in fish and aquatic invertebrates. *Environ. Toxicol. Chem.*, **1**, 347–58.

Layton, D. W., Mallon, B. J., Rosenblatt, D. H. and Small, M. J. (1987) Deriving allowable daily intakes for systemic toxicants lacking chronic toxicity data. *Reg. Toxicol. Pharmacol.*, **7,** 96–112.

Slooff, W., Van Oers, J. A. and DeZwart, D. (1986) Margins of uncertainty in ecotoxicological hazard assessment. *Environ. Toxicol. Chem.*, **5,** 841–52.

Suter, G. W. II, Vaughan, D. S. and Gardner, R. H. (1983) Risk assessment by analysis of extrapolation error, a demonstration for effects of pollutants on fish. *Environ. Toxicol. Chem.*, **2,** 369–78.

Van Straalen, N. M. and Denneman, C. A. J. (1989) Ecotoxicological evaluation of soil quality criteria. *Ecotoxicol. Environ. Saf.*, **18,** 241–51.

Zeise, L., Crouch, E. A. C. and Wilson, R. (1986) A possible relationship between toxicity and carcinogenicity. *J. Am. College Toxicol.*, **5**(2), 137–51.

# 11 *Ecological risk assessment and sustainable environmental management*

STEVEN M. BARTELL

## 11.1 INTRODUCTION

Ecological risk assessment (ERA) continues to evolve as an effective conceptual and methodological framework for estimating and evaluating the probable environmental impacts of industrial chemicals (Goldstein and Ricci, 1981; Bartell *et al.*, 1984, 1992; Bartell, 1990a,b; Suter, 1992; USEPA, 1992; Calabrese and Baldwin, 1993; Parkhurst, 1993; Pascoe, 1993). ERA, while conceived to assess ecological disturbances broadly defined, has been applied primarily in assessing the potential effects of toxic chemicals. This more narrow experience in application is understandable in that the primary legislation that mandates individual assessments derives from efforts to protect ecological resources from chemical stress. Different laws govern all phases of the life cycle of industrial chemicals, including requests to manufacture and distribute novel compounds (e.g. TSCA), commercial and agricultural practices of pesticide application (e.g. FIFRA), maintaining water quality (e.g. Clean Water Act, Ocean Dumping), and final disposal in waste sites (e.g. RCRA/CERCLA). (For a review the reader is referred to Levin and Kimball (1984) where these and other laws germane to ecological risk assessment are discussed; Bartell *et al.* (1992: Table 1.1) gives a summary listing and brief description.) Therefore, while recognizing a broader ecological justification for the development and implementation of concepts and methods for assessing ecological risk (USEPA, 1992), the following discussion focuses on ecological risks posed by toxic industrial chemicals. The major points of this chapter pertain as well to other controllable ecological

*Interconnections Between Human and Ecosystem Health.* Edited by Richard T. Di Giulio and Emily Monosson. Published in 1996 by Chapman & Hall, London. ISBN 0 412 62400 1.

disturbances (e.g. habitat destruction, declining biodiversity, genetic engineering, acid deposition, etc.).

The primary purpose of this chapter is to reinforce previous recognition (e.g. Harwell *et al.*, 1986; Suter, 1991; USEPA, 1992) of the overarching importance of selecting meaningful ecological effects in assessing ecological risk. The challenge here lies in interjecting the most current science offered by modern ecology into a decision-making process or regulatory arena where ecological principles are often poorly understood or poorly communicated, and where social, economic and political considerations enter unequally into environmental decision-making. To contribute effectively in this arena, ecological risk assessors must focus their capabilities and resources on ecological entities that are not only defensible from a scientific viewpoint, but are also of vital interest to stakeholders and decision-makers alike. Successful ERA demands the best from ecology and the best from decision-makers.

The second motivation for this chapter lies in establishing sustainable environmental management as a logical and worthwhile context for assessing ecological risks and for developing and implementing plans for environmental remediation or restoration. Integrating ecological risk assessment with sustainable management will be discussed from alternative, but complementary perspectives, that of larger scale government policy and its impact, and that of local or regional actions directed towards sustainable management. Underlying this motivation is the recognition that ERA might be developed to the point of perfection and yet ecological resources will be further degraded without a larger context within which to assess and manage ecological risks. Sustainable environmental management is that larger context.

The final reason for the chapter lies in an admonition that effective integration of ecological risk assessment and sustainable environmental management can neither guarantee nor sustain environmental quality unless meaningful plans are enacted to regulate human population pressures on finite environmental resources – at all scales. To be successful (and sustainable), environmental management must be based on vision of the desired relationship between *Homo sapiens* and the quality and quantity of environmental resources under human influence. To realize this vision, population pressures must be managed, either in shear numbers, in the per capita resource demands, or both.

The discussion begins with an overview of ecological risk assessment as currently conceived and practiced. A discussion of sustainable environmental management follows. Contributions from various environmental disciplines to an integration of ecological risk assessment with sustainable environmental management are subsequently tabulated. Institutional constraints and scale incompatibilities on acheiving such an integration are considered, along with the importance of risk-based decision-making as a common denominator for coherent environmental regulation. The chapter ends by addressing the root cause of ecological risk – namely, human population pressures that exceed ecological carrying capacities commensurate with an acceptable quality environment and an enviable quality of life.

## 11.2 ECOLOGICAL RISK ASSESSMENT

In concept, *risk* refers properly to the conditional probability of a specified event occurring (e.g. dam failure, reactor meltdown, bridge collapse, plane crash) combined with some evaluation (e.g. a loss or damage function) of the consequences of the event (e.g. injury, death, excess cancer, property loss). Kaplan and Garrick (1981) introduced a cogent conceptual model of risk as an ordered triplet (see also Helton, 1993). The first element of the triplet describes the nature of the event of interest. The probability of the event occurring designates the second element. The third member of the ordered triplet is an evaluation of the consequences of the event. By analogy, an *ecological risk* is the conditional probability of a specified ecological event occurring, coupled with some statement of its ecological consequences. As currently practiced, however, assessing ecological risk essentially entails describing, either quantitatively or qualitatively, the likely occurrence of an undesired ecological event, the consequences of which are rarely addressed. The current USEPA (1992) framework for ecological risk assessment proffers *problem definition*, *exposure assessment*, *effects assessment*, and *risk characterization* as the necessary components of an ecological risk assessment (Table 11.1). This framework extends beyond the probabilistic nature of ecological risk to include qualitative risks.

Bartell (1995) analyzed 126 abstracts summarizing case studies of ecological risk assessment. The abstracts represent 14 years (1980–1993) of ecological risk assessments reported at the annual meetings of the Society for Environmental Toxicology and Chemistry. In less than 10% of these

**Table 11.1** Components of an ecological risk assessment

| Component | Description |
|---|---|
| Problem definition | Identify disturbance (e.g. toxic chemical), delineate endpoints, suggest scope and scale of assessment |
| Exposure assessment | Describe the environmental chemistry, transport, and fate of the chemical, estimate chemical exposure profile and subsequent exposure or dose |
| Effects assessment | Define ecological effects that are consistent with the objectives of the assessment; derive the exposure–response relationships used to translate the exposure profile into risk estimates |
| Risk characterization | Integrate available information and data; estimate risks; identify and estimate uncertainties and their assessment implications |

*Source*: USEPA, 1992.

assessments was *ecological risk* defined in probabilistic terms. Nearly all of the assessments that constituted this 10% were reported in the last three years. Of historical interest, one of the first reported ecological risk assessments (presented at the 1980 meeting and later published in *Environmental Toxicology and Chemistry*, the SETAC journal) defined and estimated ecological risk as a conditional probability (O'Neill *et al.*, 1982).

Through refinement in practice and the accrual of experience, ecological assessors will advance ecological risk to conceptual and operational parity with risk as assessed in other disciplines. In the context of risk management, pressures from other components (e.g. economics, human health, engineering, technology, politics) in comprehensive risk-based decision processes (e.g. RCRA/CERCLA) will force increased sophistication in ecological risk assessment, if ecological risks are to assume a role commensurate with other risks in decision making.

### *11.2.1 SELECTING ECOLOGICAL ENDPOINTS*

*Endpoint* is a curious term in this context. Webster defines an endpoint as 'a point marking the completion of a process; esp: a point in a titration at which a definite effect . . . is observed'. In terms of the risk assessment process, selecting an endpoint is the beginning of the process, not the end. The titration analogy seems ill-fit as well; it implies a threshold in effect in relation to a continued disturbance (i.e. addition of acid or base), rather than the continuous exposure (dose)–response relationship that typically describes chemical toxicity. The relationship might exhibit a lower threshold, but the absence of measurable effects below a known threshold is of minimal interest in assessing risk. Perhaps simply using ecological *effect* or *response*, as in the effect of or the response to an ecological disturbance, might alleviate unnecessary confusion for the uninitiated. Sidestepping the terminology quagmire, however, in this chapter, endpoint, effect and response will be used interchangeably.

An endpoint in ERA therefore is an ecological effect selected as a focus for risk characterization. Chemical toxicity can be measured at different levels of biological and ecological organization. In practice, ecological responses have been defined in relation to these levels, and include effects on individuals, populations, communities, ecosystems, watersheds and landscapes. Assessments commonly address several endpoints. However, more than 90% of reported ecological assessments have focused on changes in the abundance of one or more populations as the effect of principal ecological concern (Bartell, 1995).

Attributes other than ecological contribute to identifying endpoints for ERA (Suter, 1989; USEPA, 1992). Endpoints should be consistent with the policy goals or the mandates of legislation that require the assessments. Presumably the policy and legislation reflect underlying societal values concerning ecological resources. Thus, social, political and economic attributes are important in addition to ecological considerations in selecting endpoints for risk analysis. These attributes, framed in human values, provide the link between ecological risk assessment and a broader environmental context which gives meaning to the assessment, that is, sustainable environmental management.

Selection of ecological endpoints will determine in large part the efficacy of any particular assessment because the entire process is contingent on the nature of the endpoints, which become in effect the decision criteria. The ecological specifications of endpoints selected for assessment will influence the accuracy and precision in estimating the probability of the endpoint occurring, and in evaluating the ecological consequences of an occurrence. To facilitate risk estimation, a selected ecological effect should be measurable with sufficient statistical power for a feasible investment in sampling and analysis. That is, if reference ecological measures exhibit high variance in the absence of chemical stress, extensive sampling and analysis will be required to discern chemical effects from normal variability. For example, the probability of some unacceptable decrease in the abundance of commercially valuable fish populations is a commonly selected assessment endpoint. However, large natural fluctuations in fish populations combined with limitations in accurately and precisely estimating fish population size suggest that decreased fish abundance is a questionable ecological endpoint, from the viewpoint of statistical power and risk estimation. This same concern applies to populations of birds and mammals that may carry high social or economic relevance.

The ability to evaluate the consequences of an occurrence should play an important part in defining an ecological endpoint for risk assessment. The answer to 'so what?' should be apparent and accepted by risk managers, decision-makers and other stakeholders. Surprisingly enough, justifying the selection of ecological endpoints on purely ecological grounds has proved difficult. This difficulty arises in part from an unspecified reference environment and different underlying models that might influence the ecological characteristics of such a reference (e.g. Holling, 1986). Human perceptions of the natural world will influence the definition of reference environments, selection of endpoints, and consequently, the effectiveness of any assessment. In the context of environmental regulation, the implicit model seems to be one of a constant environment and some corresponding ecological *status quo*. However, the ecological significance of measured disruption caused by chemical poisoning cannot be easily or convincingly assessed in relation to such a static model of nature. The significance of decreased productivity or even extinction cannot be judged on purely ecological grounds. In adopting the *status quo* as the frame of reference, ecological entities have no inherent ecological value; value accrues only as ecological entities are identified as resources. Nature simply is. But importantly, change is nature's nature (Holling, 1986).

The dynamics of natural systems provides a clue concerning the identification of ecologically meaningful effects for risk assessment. *Adaptability* distinguishes living systems from the inanimate (Dobzhansky, 1968). The ability to change in anticipation of change (i.e. evolution by pre-adaptation) remains the essence of living systems – populations fluctuate, yet the species may persist; species become extinct, but new species continue to evolve; community composition changes, but interactions among species remain; changing distributions of species generate complex landscape mosaics in space and time. In the face of probable extinction, life persists. The point of this hyperbole is that the ultimate ecological risk is the risk of losing the capacity to adapt. DNA is the raw material of adaptation; therefore, genetic diversity and the reproductive mechanisms that maintain, propagate, and proliferate DNA represent key foci for developing ecological risk assessment endpoints. To date, endpoints derived from even basic considerations of population genetics, fitness, or adaptability have yet to become the focus of ecological risk assessment. For example, since 1990 there have been at least two assessments that addressed potential ecological risks posed by genetically engineered microorganisms (GEMs), but the assessments addressed the potential for adverse impacts on native micro-flora and -fauna, not adaptation (Bartell, 1995). The assessments were primarily qualitative; risk was not assessed as a conditional probability. (Assessing the potential ecosystem effects of GEMs is further discussed in Orvos, 1992.) More related to risk assessment, Novak *et al.* (1985) categorized biological assays as either genetic or somatic and addressed genetic issues of cell mutations, mutations in bacteria, plants and insects, as well as changes in sister chromatid exchange in assessing the utility of these assays in 'ecoepidemiological' investigations of mixed aerosols used by the military.

Criteria for identifying ecological endpoints bear directly on the nature of the remaining components in assessing ecological risks (e.g. Table 11.1). The following briefly outlines some of the assessment implications of endpoint selection.

### *11.2.2 EXPOSURE ASSESSMENT*

The risk assessment endpoints will largely determine the quantification of exposure to toxic chemicals relevant to estimating risks. The characteristics of life history, behavior, dynamics of growth, migration and geographical distribution of selected species will specify the important spatial and temporal scales over which exposure (and dose) must be quantified. Endpoint selection may focus chemical measurements to specific environmental media, for example, soils, sediments, surface water, or air. The nature of the endpoint will determine the appropriate averaging time for exposures.

### *11.2.3 EFFECTS ASSESSMENT*

The laws of physics and chemistry determine the environmental activity of manmade chemicals; organic matter merely presents substrate and a complex array of biochemical reactions that may be blocked or altered kinetically by the presence of xenobiotics. Toxic chemicals exert their fundamental influence at the level of chemical reactions. The biological or ecological level of resolution selected to assess the impacts of altered reactions is largely a matter of the convenience of measuring and significance of the potential impact.

The challenge, stated earlier, is to introduce the best that modern ecology has to offer into the selection of endpoints. In this regard, endpoints are frequently discussed in relation to 'levels of organization'. It must be remembered that each 'level' in fact corresponds to a particular model for describing the same natural world. These levels have, for better or worse, been interpreted as a nested model of nature: landscapes that encompass ecosystems consisting of communities made of populations of individuals. Independent of the arguable utility of this simplistic model, the advances in ecological understanding achieved through decades of basic study of each different level offer unique and potentially powerful concepts and measurements that should be incorporated into the development and application of ecological risk assessment. Also, competent attempts at integrating across these levels (e.g. Allen and Starr, 1982; Allen *et al.*, 1984; Holling, 1986; O'Neill *et al.*, 1986; King, 1991) should be explored for their relevance in selecting endpoints for ERA.

### *11.2.4 RISK CHARACTERIZATION*

Risk characterization is the step in an ERA where the available information, data, models and measures are used to describe and quantify ecological risk, as defined by the selected endpoints (USEPA, 1992). Specification, quantification where possible, and examination of the implications of uncertainties that enter into the assessment are also key components of risk characterization.

An emphasis on conditional probability separates risk assessment from more traditional environmental impact assessment. The explicit incorporation and carrying through of uncertainty in estimating risks further distinguishes risk assessment from impact assessment. Despite current acceptance of qualitative statements of risk, the drive for common parlance among risk assessors in allied disciplines (e.g. hydrology, engineering, process safety) will force ecological risk characterization towards a more probabilistic foundation. This is not to deny that scientifically defensible and ecologically beneficial decisions can (and indeed do) result from qualitative assessment of ecological impacts. The point is to reserve characterization of ecological *risk* for probabilistic statements consistent with the Kaplan and Garrick (1981) ordered triplet model for risk and to avoid simply conferring a new title to traditional environmental impact assessment.

The emphasis on the probabilistic nature of risk carries implications for methods acceptable for characterizing ecological risk. One implication of adopting the conditional probability model for risk is that the often used quotient method does not characterize risk. (Quotients are calculated by dividing a measured or estimated exposure concentration of toxic chemical by a concentration suspected to cause an undesired ecological effect, for example, mortality, chronic sublethal effects, or reproductive effects. An application factor may be used to modify the toxicity concentration, depending on the data source. Quotients$>1$ have been interpreted as designating potential ecological risk; quotients$<1$ may indicate minimal risk. Others such as Cardwell *et al.* (1993) have used values of 0.3 as threshold values for assessing potential risks.) Quotients are not constrained to the same range of values as risk, e.g. $\{0 < P<1\}$, where $P$ designates the probability of occurrence for an endpoint. Quotients are none the less valuable in screening calculations that can reduce the dimensionality of a subsequent ecological risk assessment (e.g. Cardwell *et al.*, 1993; Bartell and Wittrup, 1995). The important point is that methods used to estimate a *probability* of an ecological effect differ from methods used to quantify the effect. Endpoints for risk should be chosen mindful of such differences.

Similar deliberations apply to other methods used to characterize ecological risks. Experiments, performed under either laboratory or field conditions, aimed at assessing risks must be designed with the necessary replication to estimate a probability of an effect, not simply an effect, occurring in relation to the disturbance, e.g. toxic chemicals. Serious attempts to experimentally quantify risks will clearly be costly.

Using mathematical models to assess risk, model validation questions aside, requires careful deliberation on the nature of statistical uncertainties propagated through the models and the methods of propagation. Different approaches (e.g. Monte Carlo methods, first order variance propagation, fuzzy arithmetic) can be equally justified in terms of mathematical rigor. However, these methods can produce different model results (and different risk estimates) using model inputs derived from the same background information (Scavia *et al.*, 1981).

### *11.2.5 WHAT ENVIRONMENT?*

If ecological risk assessment is to realize its full potential in characterizing probable human impacts on the environment, a clearer description of 'environment' is required. Throughout the development of ERA, there has been an implicit reference to 'the natural past' (i.e. Power-Bratton, 1992) as one underlying model for 'the environment' in the broader sense. This model (or any other, e.g. Holling, 1986) reference environment has yet to be described with sufficient rigor to provide clues for delineating meaningful assessment endpoints. In response to the more local demands of RCRA/CERCLA baseline assessments, the absence of a clear definition of environmental quality is evidenced in the criteria used to select reference sites to compare with sites potentially impacted by chemical wastes (e.g. East Fork Poplar Creek study, USDOE, 1994). Lacking a prescription for the kind (i.e. quality, quantity) of environment that is the goal of protective mandates, risk assessors cannot make best use of their ecological or quantitative skills. Ironically, the same general wording of environmental legislation that anticipates the need for flexibility in judicial interpretation runs counter to developing the desired rigor, precision and scientific justification of endpoints used to enforce the legislation.

Legislation that currently drives the majority of site-specific ecological risk assessment pertains, by definition, to relatively small spatial scales. Within these smaller scales, it is possible through careful selection of ecological endpoints to develop the capability to assess risk accurately and precisely. In a broader environmental context, effective smaller scale assessments may still translate into continued environmental degradation at larger scales, that is, risk assessments that meet RCRA/CERCLA compliance criteria may be irrelevant to environmental quality in areas adjacent to waste sites. Considerations of a larger scale context for risk assessment might stimulate novel solutions to waste site remediation and facilitate taking advantage of any legislative flexibility in community compensation for accepting environmental (or human health) risks.

The remainder of this exposition promotes one potential model for environmental quality that might provide the broader scale context for more productive development and implementation of ecological risk assessment. That model is sustainable environmental management.

## 11.3 SUSTAINABLE ENVIRONMENTAL MANAGEMENT

Simply stated, sustainable environmental management maintains the quality and quantity of ecological resources at approximate steady-state. Resources are depleted at rates not to exceed rates of their renewal. Depletion rate is measured over scales in time and space relevant to the resource and the user. Clearly, ecological resources cannot be sustainably managed without a prescription regarding quantity and quality. Thus, embracing *sustainable environmental management* as an overarching management objective, decision-makers and risk managers are forced to delineate, in operational detail, the currently missing underlying environmental model. (It can be noted in passing that in August 1993, more than 40 ecologists, policy analysts, risk assessors, lawyers, representatives from industry and government officials participated in a working conference that addressed the potential integration of risk analysis and sustainable environmental management. The conference was held at the University of Michigan Biological Station at Pellston, Michigan, under the auspices of the Society of Environmental Toxicology and Chemistry and the Ecological Society of America.)

## 11.4 TOP-DOWN AND BOTTOM-UP APPROACHES TO SUSTAINABILITY

Two approaches contribute towards developing and implementing sustainable environmental management practices. Strategic national and international discussions, commitments and treaties or protocols can provide a foundation for future policy or regulations that, in spirit, stimulate change from the 'top down'. Such larger scale institutional actions encourage or compel state and local reaction.

Complementary to the incentives and policies promulgated by larger scale institutions are the 'grass roots' actions of local and regional groups who seek to develop and implement sustainable environmental management plans without the insistence (or interference) of 'big government'. The tactic of these smaller scale movements is to force larger scale changes from the bottom up. Both 'bottom-up' and 'top-down' approaches need to be effectively combined to integrate risk assessment with sustainable environmental management.

### *11.4.1 TOP-DOWN APPROACHES*

Top-down approaches to risk management are certainly important and effective for managing resources that have characteristic spatial scales which exceed the jurisdiction of local or regional institutions. When local management activities impact upon resource availability at larger scales, institutions with jurisdiction at these larger scales are necessary to ensure equitable use and sustainability of the resource. For example, marine fisheries resources must be managed to guarantee sustained access by fishermen from many states. Equitable distribution of water rights in the arid western and southwestern United States remains a volatile political and economic issue requiring federal intervention. Other examples of top-down policies and activities that are consistent with the aims of sustainable environmental management include signing of the Montreal Protocols to decrease greenhouse gas emissions, participation in the 1991 Earth Summit meeting at Rio de Janeiro, and the recent formation of the President's council on sustainable development.

At the level of state government, Governor Arne Carlson initiated Minnesota's efforts to promote sustainable development in January 1993. At a February 1994 workshop, 105 leaders from business and environmental organizations outlined strategies to 'redefine the relationship between economic activity and the natural world'. Of course, top-down approaches to sustainability are not limited to governments. In competition for designing the site of the next world's fair to be held in Hanover, Germany, potential design firms have been asked to develop their plans in accordance with the nine Hanover Principles for Sustainable Development (William McDonough Architects, 1992).

One advantage of the top-down approach is that resulting policy or legislation, once implemented, can have large scale impact in a relatively short time. One drawback is that large-scale policy may be inappropriate at the scale of impact. For example, regional differences in geochemistry are not included in the development of national water quality criteria; differences in water chemistry might result in standards too restrictive in some watersheds or not protective in others. Another drawback is the often lengthy process required to enact laws or develop policy at these larger institutional scales.

### *11.4.2 BOTTOM-UP APPROACHES*

Regional differences in the distribution of valued ecological resources suggest that more localized approaches to sustainable environmental management may prove more effective and more efficient than broad sweeping top-down policies in addressing environmental risks.

Regional planning has historically been an effective bottom-up approach to sustainable environmental management which itself has a historical precedence in North America to the current economic model which emphasizes personal consumption and resource depletion (Sargent *et al.*, 1991). Environmental planning on the North American continent predates European colonization of the eastern seaboard by several centuries. The tenth-century Hohokam, the Anasazi, and other Pueblo Indian cultures in the arid Southwest developed sophisticated agricultural communities based on complex irrigation systems. Village designs of the Anasazi of Chaco Canyon featured architecture and siting of structures that facilitated passive solar heating.

The layout of Spanish settlements in America was governed by the 148 separate ordinances that constituted the Law of the Indies, as issued by King Philip II in 1573. These laws dictated the careful siting of new settlements in relation to water supply, topography, soil fertility and local ecological resources suitable for crops, livestock, orchards, etc. City planning in relation to these ordinances produced the gridiron system of plazas that characterize early Spanish cities as diverse as Santa Fe, Los Angeles and St Louis.

In contrast, settlement in New England reflected a frontier ethic where acreage was cleared and defended by a diversity of individual interests including trapping, logging, mining, farming, and land speculation. Planning was minimal, limited primarily to the layout and construction of roads connecting settlements for convenience of local militia; land development was determined by powerful individuals or small corporations. The circumstances of weak state governments, minimal planning and strong local governments were replaced by significant federal involvement in the 1930s, when the Great Depression forced financially strapped state governments to embrace projects involving navigation, hydroelectric power, flood control, soil protection, national forests, scenic parkways and commercial agriculture.

The early 1920s saw the establishment of the Regional Planning Association of America. The RPAA formed in response to concern over aggressive logging and mining in rural regions combined with uncontrolled growth in industrial centers (Sargent *et al.*, 1991). The RPAA reflected the strong regionalism of the early twentieth century United States, and viewed regional entities as distinctive combinations of geography, economic and cultural elements. An emphasis of the RPAA was a balance between rural areas and metropolitan centers, within and among regions. Key issues in regional planning included environmental quality, public involvement, and the equitable distribution of wealth and opportunity.

The fundamental values and concepts of RPAA date to the utopian visions of Robert Owen and Ebenezar Howard (e.g. garden cities). Other prominent individuals include forrester Benton MacKaye, one of the founders of the RPAA; Louis Mumford, who redefined the planning process to include community, as well as professional, participation; and a southern sociologist, Howard Odum Sr., who emphasized community education and ecological balance as prime ingredients of regionalism. The subsequent roles of Odum's sons in establishing the science of ecology attests to the close relationship between ecology, related environmental disciplines, and regional planning.

The components of regional planning (Table 11.2) lend themselves to integration with the objectives and components of ecological risk assessment.

At a more local scale, private developers, architects, contractors and builders are using innovative designs, natural or non-toxic materials and construction practices that are consistent with environmental sustainability. Somewhat surprisingly, most practicing professional ecologists remain unaware of this activity. For example, the American Institute of Architects recently sponsored a three-part teleconference series that addressed critical issues concerning sustainable development (AIA, 1993a,b,c). The operational philosophy of the AIA regarding sustainability is essentially 'sustainable development, . . . one building at a time'. These developers recognize that

**Table 11.2** Components of the regional planning process

| Component of planning | Relevance to ecological risk assessment |
|---|---|
| Discover public goals | Assist in problem formulation, resource evaluation, risk ranking |
| Inventory the resource base | Endpoint selection, problem formulation |
| Protect natural areas | Problem formulation, resource evaluation |
| Maintain land in agriculture | Landscape level risk assessment, endpoint selection, exposure analysis |
| Maintain water resources | Endpoint selection, exposure analysis |
| Preserve recreational, historic sites | Problem formulation, define scales of assessment |
| Increase and share equity | Evaluate consequences of risk, risk ranking |

*Source*: Sargent *et al.* (1991).

technological advances cannot be expected to solve the fundamental problems leading to unsustainable growth and development. However, these 'activists' demonstrate the ability to take advantage of the partial solutions offered by technologically sophisticated building design and construction, ecologically sensible building materials, and facility designs that produce structures that are integrated with the local landscape, rather than merely imposed upon it.

One noteworthy example of the combined effects of top-down and bottom-up approaches to sustainable development is the increase in energy efficient building design and construction by private contractors and pending federal legislation that will require refitting of approximately 500 000 federal government buildings to increase energy efficiency (AIA, 1993b). Clearly, the necessary federal funds to implement such legislation will be difficult to appropriate given present and projected budget constraints. Nevertheless, the incentives provided by the spirit of the legislation may continue to spur the sustainable development activities in the private sector.

A variety of environmental disciplines can contribute to integrating ecological risk assessment with sustainable environmental management (Table 11.3). Concepts and methods borrowed from landscape ecology, wildlife management, ecological restoration and ecological engineering should be incorporated into consideration of waste site remediation. These disciplines can assist in defining meaningful assessment endpoints and in developing remediation goals that put remediation in a spatial context extending beyond the waste site boundaries. Ecological economics might stimulate the new and innovative relationships among ecological, legal, political and social institutions that are required to implement plans for sustainable management.

## 11.5 LANDSCAPE ECOLOGY AND HUMAN DOMINATED SYSTEMS

Ecological risk assessments address potential problems that are inherently spatial, as well as temporal in scale. Regardless of the nature of the ecological disturbance or the ecological endpoint(s) of interest, each ecological risk assessment implies some characteristic space–time frame. Furthermore, industrial chemicals are introduced to the environment from point and non-point sources that occupy locations on landscapes. Therefore, from the perspectives of ecological resources at risk and the source of chemical disturbances, landscape ecology stands to contribute fundamentally to the future development of ecological risk assessment.

Landscape ecologists have come to recognize the close coupling between human activities and ecological pattern and process on the landscape. Several plenary speakers at the 1993 annual meeting of the International Association of Landscape Ecologists, held in Oak Ridge, Tennessee, called for the explicit representation and linkage of human activities in the development of landscape models for exploring and addressing ecological disturbances. This recognition of the relevance of humans to shaping ecological relationships contrasts with traditional perspectives in basic ecological research, which has focused historically on understanding pristine ecological systems, or pretending to ignore the impacts of humans. Blennow and Hammarlund (1993) describe the impacts of nineteenth century societal decisions concerning reafforestation on current forest vegetation and dynamics in Halland, Sweden. This case study is an excellent example of the interplay between human activities and ecological processes in determining the nature and distribution of resources on dynamic landscapes. Their work also raises the important point regarding preservation *vs.* conservation of natural resources.

Landscape ecologists continue to explore the propagation of disturbances (natural and man-made) across landscapes (e.g. Turner, 1989; Rykiel *et al.*, 1988; Turner and Dale, 1991). Indeed, ecological risk assessment might be considered a subset of general ecological disturbance theory (e.g. Pickett and White, 1985). Two aspects of landscape ecology promise to contribute to developing risk assessment capabilities at scales relevant to point source and non-point chemical stress. First, work continues in describing and quantifying patterns of resource distribution on landscapes (Krummel *et al.*, 1987; O'Neill *et al.*, 1988). Secondly, complementary work proceeds in hope of identifying and quantifying the critical underlying environmental processes (including human activities) that determine patterns on landscapes (Turner, 1989).

Similar discussions can be developed for the remaining disciplines listed in Table 11.3. All offer unique, but interrelated concepts and methods to increase current capabilities in assessing ecological risks and to facilitate the necessary integration of ERA with sustainable environmental management.

## 11.6 REMEDIATION, RESTORATION AND SUSTAINABLE ENVIRONMENTAL MANAGEMENT

The concept of risk-based remediation or restoration is consistent with local and regional planning based on sustainable development. Adopting this concept forces remediation and restoration objectives to be compatible with environmental risks associated with current and planned land use. Remediation or restoration based on criteria or standards may force clean-up efforts to exceed rational expectations in some cases, or may be ineffective in protecting ecological resources in other instances. For example, technological advances in analytical chemistry have decreased the concentrations of many chemicals that can be detected in samples of air, soils, sediments, water and organisms. Given the high degree of uncertainty associated with the development of chemical criteria, the temptation exists to impose more stringent standards as detection levels decrease. The net result can be costly and excessive clean-up activities that provide ever-diminishing returns in reduced risk and environmental protection per additional dollars spent.

Risk-based remediation provides the framework for local public involvement

**Table 11.3** Potential contributions from different environmental sciences to ecological risk assessment in the context of sustainable environmental management

| Environmental discipline | Contribution to sustainable management |
|---|---|
| Landscape ecology (Turner and Gardner, 1991) | Describe pattern and process on landscapes, relate human activities to ecological impacts |
| Ecological restoration (Berger, 1990; Mollison, 1990; Kidd and Pimentel, 1992; Harker *et al.*, 1993) | Provide methods for remediating waste sites in proper ecological context with surrounding landscape, identify species assemblages, link restoration to ecological engineering |
| Wildlife management (Adams and Leedy, 1991; Rodiek and Bolen, 1991; Starfield and Bleloch, 1991) | Further guide ecological restoration to provide habitat and resources to sustain fauna that immigrate or are introduced to the former waste site |
| Ecological engineering (Johanasson *et al.*, 1993) | Link with restoration ecology to implement designs with nature |
| Ecological economics (Costanza, 1991; Cairncross, 1992; Costanza *et al.*, 1992; Peet, 1992) | Realistic valuation of ecological resources used in costing consequences of ecological risks and estimating costs of remediation and restoration |

in establishing acceptable levels of risk in relation to other activities that determine current or planned environmental quality in the region. Remediation of a waste site to detection-level concentrations may prove an unwise investment if the adjacent unremediated lands pose unacceptable risks to ecological resources. Cleaning up a facility to background environmental concentrations of potentially toxic chemicals may simply expend limited resources if the facility will serve a similar purpose in the future. For example, how stringently should a former air force or naval base be remediated if it will become a municipal airport or commercial port upon completion of the clean-up? If consensus concerning acceptable risks can be obtained for the future facility, current clean-up efforts can target a similar level of risk. This perspective does not mean to imply that in some cases a risk-based remediation might not insist on more stringent clean-up than federal or state standards. However, risk provides a common denominator for assessment. In contrast, more stringent standards might not correspondingly reduce risk.

## 11.7 DISPARATE SCALES IN ECOLOGICAL AND INSTITUTIONAL DYNAMICS

One difficulty in effectively addressing ecological concerns or intelligently managing ecological resources lies in the scale incompatibilities of ecological phenomena and human institutions entrusted with their stewardship. Ecological resources exhibit a diversity of characteristic temporal scales. Diurnal and lunar cycles are important to the growth and survival of some species. The life histories or production dynamics of many species are strongly seasonal. Yet for others, individual lives span decades, even centuries.

Businesses, banks, corporations and other economic entities are closely tied to quarterly reports and annual financial statements. Government institutions are active in relation to their fiscal budgeting cycles, as well as the two- and four-year cycles associated with elections. Legislative institutions can introduce significant time lags into the management process.

Additionally, Holling (1986) observes that political institutions exhibit another temporal scale with a periodicity of 20–30 years. This period corresponds roughly to the turnover time of a generation of policy-makers or regulators. This turnover permits re-evaluation of past policies and decisions by the following generation of politicos. Decisions can be overturned and policy revised with minimal ill-will among personnel.

Appreciable differences in the activity patterns of regulatory or policy-making institutions and the ecological resources entrusted to or influenced by these institutions can pose nearly insurmountable problems in designing and implementing effective policy. One answer to the scale discrepancies is to permit the relevant time scale of the resource of concern to dictate the scale of the management activity. The Iroquois, for example, 'planned for the seventh generation' (Sargent *et al.*, 1991), which approximates the 250-year life span of the dominant trees in the forests that provided resources to this nation. Lacking a significant energy subsidy in the form of fossil fuels, these natives were more closely constrained by the natural dynamics of ecological resources than are members of fossil-fueled industrial society. Therefore, it is not surprising that the 'management institutions' of the Iroquois were more closely scaled to the dynamics of their ecological life-support systems.

A Carnegie Commission report of 1993 called for increased cooperation among federal agencies entrusted and empowered to maintain environmental quality. Risk assessment can provide a common operational framework to facilitate communication and cooperation. Transcending agency boundaries and jurisdictions can markedly and efficiently impact environmental quality. If the Department of Transportation lobbied for and enforced increases in fuel efficiency that are technologically possible, jobs at the US EPA offices responsible for air quality may become much easier. The political will to re-invent government to increase operational and scale compatibility with twenty-first century problems remains crucial to achieving sustainability at a degree of environmental quality worth preserving.

Relocating the power to make decisions to local and regional institutions appears as another step towards reducing scale incompatibilities. As components of an integrated local or regional plan, ecological resources might be more effectively managed by those groups who know and understand the spatial–temporal dynamics of the resource in question. Instead of broad sweeping management aimed at average circumstances (which might not exist), management can be customized for the particular nature and distribution of resources that define regional differences.

## 11.8 THE POPULATION BOMB – 1990s

The explosion continues. From 1800 to 1992 the world human population expanded by nearly five-fold – from approximately 1 billion people to 5.5 billion. Combined with the global population growing at an exponential rate is the recognition that individuals in industrialized societies have increased their per capita demands on energy and ecological resources. The net result continues as unsustainable and unprecedented human pressures on global resources (Ehrlich and Ehrlich, 1992). Even in 'developed' societies with stable or decreasing populations, the ecological implications of economic requirements for continuous growth and expansion are manifestly inconsistent with sustainability. Earth's inhabitants are participating, willingly or not, in an uncontrolled experiment based on ever-intensive exploitation of diminishing planetary resources. The environmental impacts of this experiment are increasingly evident, now even at global scales as loss of valuable habitat, loss of prime agricultural lands, increased atmospheric $CO_2$ concentrations, and decreases in the protective stratospheric ozone concentrations.

Hierarchy theory (Allen and Starr, 1982; O'Neill *et al.*, 1986) warns of the implications of overconnected systems. In the absence of human-induced controls, natural constraints on continuous growth (e.g. loss of top soil, increased desertification, social polarization and revolution) will impose controls that may result in overall system collapse to a new and different hierarchical structure. *Homo sapiens* might or might not be part of that new order. The choice is one of self-imposed constraints on population size, or perhaps more severe constraints imposed by nature.

Consideration of population density and distribution has direct bearing on integrating ecological risk assessment with sustainable environmental management. Such integration will be more effective under circumstances where community participants have a clear vision of their desired relationship to ecological resources that they influence and in turn are influenced by. Part of this vision must include consideration of the local or regional relationship between a sustainable, desirable quality of life, as defined by the community, and the number of people that can be supported at any selected level of quality. Development of ecological risk assessment or sustainable environmental management that fails to meaningfully address human population dynamics will be as effective as efforts to cure the auto-immune deficiency syndrome (AIDS) that ignored the human immunodeficency viruses (HIV).

Integrating ecological risk assessment with sustainable environmental management, by focusing on relations involving population, planning, sustainability and risk offers an important step towards addressing causes instead of merely assessing symptoms.

## 11.9 IN THE END

The continued refinement and improvement of concepts and methods used to assess ecological risks are necessary and justifiable for complying with legislation designed to protect the environment from the impacts of toxic chemicals. However, capabilities for assessing ecological risks will be developed fully and applied with the greatest societal benefit only when ERA becomes an integrated component of an overall plan for sustainable environmental management.

In the absence of a larger, more comprehensive environmental context, ERA might be developed to the point of completely accurate risk estimation and flawless assessment in relation to environmental legislation, yet the net result could be continued degradation and loss of ecological resources and human life-support systems.

To be effective in the short term, planning for sustainable environmental management will likely be mainly local or regional in scale. Local planning and implementation can reduce some of the incompatibilities between the relevant space–time scales of ecological systems and the scales of political and economic activity needed to restore and sustain these systems into the future.

Ecological resources must be represented explicitly and realistically in economic models used for local or regional planning. Such omission may have been justifiable when human population sizes were comparatively small, energy subsidies to human activities were minimal, and ecological resources appeared limitless. Such omission is no longer acceptable in developing credible economic models. The past unrealistic accounting of ecological resources is one reason why the research, development and assessment activities of the kind reported throughout this book have been necessary.

Assessment endpoints in ERA address only the symptoms of a rapidly expanding population of human beings. Failure to meaningfully address problems of population density, distribution and demands on resources (i.e. lifestyle) will only sustain the need for more assessments and put into action the global feedback mechanisms for balancing human population demands with supplies of renewable resources. This is perhaps the greatest ecological risk.

## REFERENCES

Adams, L. W. and Leedy, D. L. (1991) *Wildlife Conservation in Metropolitan Environments*, National Institute for Urban Wildlife, Columbia, MD.

AIA (American Institute of Architects) (1993a) *Building Connections. Program I: Energy and Resource Efficiencies*, AIA, Washington, DC.

AIA (American Institute of Architects) (1993b) *Building Connections. Program II: Health Buildings and Materials*, AIA, Washington, DC.

AIA (American Institute of Architects) (1993c) *Building Connections. Program III: Land, Resources, and Urban Ecology*, AIA, Washington, DC.

Allen, T. F. H. and Starr, T. B. (1982) *Hierarchy: Perspectives for Ecological Complexity*, University of Chicago Press, Chicago, IL.

Allen, T. F. H., O'Neill, R. V. and Hoekstra, T. W. (1984) *Interlevel Relations in Ecological Research and Management: Some Working Principles from Hierarchy Theory*, USDA Forest Service General Technical Report RM-110. Rocky Mountain Forest and Range Experiment Station, Fort Collins, CO.

Bartell, S. M. (1990a). Ecosystem context for estimating stress-induced reductions in fish populations. *American Fisheries Society Symposium* No. **8,** pp. 167–82

Bartell, S. M. (1990b). Issues of rate structure, scaling, and aggregation in ecological risk analysis, in *Risk-based Decision Making in Water Resources* (eds Y. Y. Haimes and E. Z. Stakiv), ASCE, New York.

Bartell, S. M. (1995) Ecological/environmental risk assessment, in *Risk Assessment and Management Handbook for Environmental, Health, and Safety Professionals*, (R. Kolluru, S. Bartell, R. Pitbaldo and S. Stricoff), McGraw-Hill Inc., New York.

Bartell, S. M. and Wittrup, M. (1995) McArthur River ecological risk assessment, in *Risk Assessment and Management Handbook for Environmental Health and Safety Professionals*, (R. Kolluru, S. Bartell, R. Pitbaldo and S. Stricoff), McGraw-Hill Inc., New York.

Bartell, S. M., Gardner, R. H. and O'Neill, R. V. (1992) *Ecological Risk Estimation*, Lewis Publishers, Chelsea, MI.

Bartell, S. M., O'Neill, R. V. and Gardner, R. H. (1984) Modeling effects of toxicants on aquatic systems, in *Water for Resources Development* (ed. D. L. Schreiber), American Society of Civil Engineers, New York.

Berger, J. J. (1990) *Environmental Restoration*, Island Press, Washington, DC.

Blennow, K. and Hammarlund, K. (1993) From heath to forest: land-use transformation in Halland, Sweden. *Ambio*, **22,** 561–7.

Cardwell, R. D., Parkhurst, B., Warrin-Hicks, W. and Volosin, J. (1993) Aquatic ecological risk assessment and clean-up goals for metals arising from mining operations, in *Applications of Ecological Risk Assessment to Hazardous Waste Site Remediation* (eds E. S. Bender and F. A. Jones), Water Environment Federation, Alexandria, VA, pp. 61–72.

Cairncross, F. (1992) *Costing the Earth*, Harvard Business School Press, Boston, MA.

Calabrese, E. J. and Baldwin, L. A. (1993) *Performing Ecological Risk Assessments*, Lewis Publishers, Chelsea, MI.

Carnegie Commission (1993) *Risk and the Environment. Improving Regulatory Decision Making*. A Report of the Carnegie Commission on Science, Technology and Government. Carnegie Commission, New York.

Costanza, R. (1991) *Ecological Economics*, Columbia University Press, New York.

Costanza, R., Norton, B. G. and Haskell, B. D. (eds) (1992) *Ecosystem Health*, Island Press, Washington DC.

Dobzhansky, T. (1968) Adaptedness and fitness, in *Population Biology and Evolution* (ed. R. C. Lewontin), Syracuse University Press, Syracuse, NY, pp. 109–21.

Ehrlich, A. H. and Ehrlich, P. R. (1992) Ecosystem risks associated with the population explosion, in *Predicting Ecosystem Risk* (eds J. Cairns, Jr, B. R. Niederlehner and D. R. Orvos), Princeton Scientific Publishing, Princeton, NJ, pp. 9–21.

Goldstein, R. A. and Ricci, P. F. (1981) Ecological risk uncertainty analysis, in *Energy and Ecological Modeling* (eds W. J. Mitsch, R. W. Bosserman and J. M. Klopatek), Elsevier, Amsterdam.

Harker, D. *et al.* (1993) *Landscape Restoration Handbook*, CRC Press, Boca Raton, FL.

Harwell, M. A., Harwell, C. C., Weinstein, D. A. and Kelly, J. R. (1986) *Anthropogenic Stresses on Ecosystems: Issues and Indicators of Response and Recovery*. ERC-153, Ecosystems Research Center, Cornell University, Ithaca, NY.

Helton, J. C. (1993) Risk, uncertainty in risk, and the EPA release limits for radioactive waste disposal. *Nuclear Technology*, **101,** 18–39.

Holling, C. S. (1986) The resilience of terrestrial ecosystems: local surprise and global change, in *Sustainable Development of the Biosphere* (eds W. C. Clark and R. E. Munn), International Institute for Applied Systems Analysis, Laxenburg, Austria.

Johanasson, T. B. *et al.* (1993) *Renewable Energy*, Island Press, Washington, DC.

Kaplan, S. and Garrick, B. J. (1981) On the quantitative definition of risk. *Risk Analysis*, **1,** 1.

Kidd, C. V. and Pimentel, D. (1992) *Integrated Resource Management*, Academic Press, San Diego.

King, A. W. (1991) Translating models across scales in the landscape, in *Quantitative Methods in Landscape Ecology* (eds M. G. Turner and R. H. Gardner), Springer-Verlag, NY, pp. 479–517.

Krummel, J. R., Gardner, R. H., Sugihara, G. *et al.* (1987) Landscape patterns in a disturbed environment. *Oikos*, **48,** 321–4.

Levin, S. A. and Kimball, K. D. (eds) (1984) New perspectives in ecotoxicology. *Environmental Management*, **8,** 375–442.

Mollison, B. (1990) *Permaculture. A Practical Guide for a Sustainable Future*, Island Press, Washington, DC.

Novak, E. W., Porcella, D. B., Johnson, K. M. *et al.* (1985) Selection of test methods to assess ecological effects of mixed aerosols. *Ecotoxicology and Environmental Safety*, **10,** 361–81.

O'Neill, R. V., Gardner, R. H., Barnthouse, G. W. *et al.* (1982) Ecosystem risk analysis: a new methodology. *Environmental Toxicology* and *Chemistry*, **1,** 167–77.

O'Neill, R. V., DeAngelis, D. L., Waide, J. B. and Allen, T. F. H. (1986) *A Hierarchical Concept of Ecosystems*, Princeton University Press, Princeton, NJ.

O'Neill, R. V. *et al.* (1988) Indices of landscape pattern. *Landscape Ecology*, **1,** 153–62.

Orvos, D. R. (1992) Assessing environmental risk from genetic engineered microorganisms and products containing recombinant DNA, in *Predicting Ecosystem Risk*, (eds J. Cairns, Jr, B. R. Niederlehner and D. R. Orvos), Princeton Scientific Publishing, Princeton, NJ, pp. 215–35.

Parkhurst, B. (1993) Framework for ecological risk assessment, in *Applications of Ecological Risk Assessment to Hazardous Waste Site Remediation*. (eds E.S. Bender and F.A. Jones), Water Environment Federation, Alexandria, VA.

Pascoe, G. A. (1993) Wetland risk assessment. *Environmental Toxicology and Chemistry*, **12,** 2293–307.

Peet, J. (1992) *Energy and the Ecological Economics of Sustainability*, Island Press, Washington, DC.

Pickett, S. T. A. and White, P. S. (1985) *The Ecology of Natural Disturbance and Patch Dynamics*, Academic Press, NY.

Power-Bratton, S. (1992) Alternative models of ecosystem restoration, in *Ecosystem Health* (eds R. Costanza, B. G. Norton and B. D. Haskell), Island Press, Washington, DC, pp. 170–89.

Rodiek, J. E. and Bolen, E. G. (1991) *Wildlife and Habitats in Managed Landscapes*, Island Press, Washington, DC.

Rykiel, E. J., Coulson, R. N., Sharpe, P. J. H. *et al.* (1988) Disturbance propagation by bark beetles as an episodic landscape phenomenon. *Landscape Ecology*, **1,** 129–39.

Sargent, F. O. *et al.* (1991) *Rural Environmental Planning for Sustainable Communities*, Island Press, Washington, DC.

Scavia, D., Powers, W. F., Canale, R. P. and Moody, J. L. (1981) Comparison of first-order error analysis and Monte Carlo simulation in time-dependent lake eutrophication models. *Water Resources Research*, **17,** 1051–9.

Starfield, A. M. and Bleloch, A. L. (1991) *Building Models for Conservation and Wildlife Management*, Burgess International Group, Edina, MN.

Suter, G. W. II (1989) Ecological endpoints, in *Ecological Assessment of Hazardous Waste Sites: a Field and Laboratory Reference Document* (eds W. Warren-Hicks, B. R. Parkhurst and S. S. Baker, Jr), US EPA 600/3–89/013, Washington, DC.

Suter, G. W. II (1991) *Screening Level Risk Assessment for Off-site Ecological Effects in Surface Waters Downstream from the US Department of Energy Oak Ridge Reservation.* ORNL/ER-8, Oak Ridge, TN.

Suter, G. W. II (ed.) (1992) *Ecological Risk Assessment*, Lewis Publishers, Chelsea, MI.

Turner, M. G. (1989) Landscape ecology: the effect of pattern on process. *Annual Reviews of Ecology and Systematics*, **20,** 171–97.

Turner, M. G. and Dale, V. H. (1991) Modeling landscape disturbance, in *Quantitative Methods in Landscape Ecology* (eds M. G. Turner and R. H. Gardner), Springer-Verlag, NY, pp. 323–51.

Turner, M. G. and Gardner, R. H. (eds) (1991) *Quantitative Methods in Landscape Ecology*, Springer-Verlag, NY.

USDOE (1994) Feasibility Study for the East Fork Poplar Creek Sewer Line Beltway. DOE/OR/02–1185 & D2 & VR. US Department of Energy, Oak Ridge Operations Office, Oak Ridge, TN.

USEPA (1992) *Framework for Ecological Risk Assessment*, EPA/630/R-92/001, US Environmental Protection Agency, Washington, DC.

William McDonough Architects (1992) *The Hannover Principles: Design for Sustainability*, Plowshares Press, Princeton, NJ.

# PART FIVE

## *Socioeconomic and Psychological Perspectives*

# 12 Measuring economic values for ecosystems

V. KERRY SMITH

To the average person the term *value* means importance or desirability. This concept seems so straightforward as to raise questions about why there could possibly be disputes between ecologists, economists and philosophers over the value of ecological resources.[1] The problems arise when we attempt to make this general definition operational. To do so requires adopting the perspective of a discipline. Value for economists must derive from anthropogenic roots. That is, people's preferences provide the sole basis for the economic values associated with all types of resources. Unfortunately, this position is often misconstrued. It does not imply that values are exclusively associated with private commodities exchanging on markets.[2] Rather the focus of the definition should be on **people** and their judgments about what is important.

None the less, it would not be fair to those criticizing the economist's approach to valuation to leave out consideration of the important issues arising in how economists have used their framework. What is learned from any model, including the economic models used to describe people's decisions, depends on how the framework characterizes individuals' preferences and measures commodities. These judgments provide the basis for all monetary measures of people's values for any good or service. The next two sections of this chapter outline some of the issues involved in describing how ecosystems

[1] For a discussion of the contrasting views, see the exchange between Kopp (1993) and Sagoff (1993) in Resources for the Future's *Resources* as well as the summary developed as part of the Conservation Foundation's Ecosystem Valuation Forum, *Issues in Ecosystem Valuation: Improving Information for Decision Making*, Phase I report to US Environmental Protection Agency Resolve, 1992.

[2] Hanemann's (1994) recent overview of the evaluation of the new economic paradigm that identifies individual willingness to pay and not prices as the primary basis for evaluating allocation decisions describes why this has been so important for environmental resources.

*Interconnections Between Human and Ecosystem Health.* Edited by Richard T. Di Giulio and Emily Monosson. Published in 1996 by Chapman & Hall, London. ISBN 0 412 62400 1.

are represented as economic commodities and how they are hypothesized to influence people. Before turning to this discussion, it is also important to consider why increasing attention has been given to the interrelationship between human activity and the health of ecosystems at both regional and global scales.

## 12.1 ECOSYSTEM VALUATION IN PERSPECTIVE

Today's concerns about the importance of ecological resources are the result of a progressive evolution in the thinking of natural and social scientists about the interactions between human activity and environmental resources. To appreciate current interest in gauging the economic value of ecological resources, it is important to consider, briefly, the context that has brought us to this point. Natural resource policy in the US has experienced periodic episodes where national attention focused on the relationship between the availability of these resources and the ability of our economy to sustain a high level of material well-being (for review, see Smith, 1980). While similar periods of concern could also be identified throughout recorded history, the most recent episodes have displayed a new set of issues that bear directly on the valuation of ecological resources. A slow transition to this new orientation, emphasizing non-market resources, began with analyses motivated by the Paley Commission report. For its authors, the interaction between natural resources and people was identified at that time as the 'Materials Problems' (see President's Materials Policy Commission, 1987). The problem was described as arising because:

> the consumption of almost all materials is expanding at compound rates and is thus pressing harder and harder against resources which, whatever else they may be doing, are not similarly expanding. This materials problem is thus not the sort of 'shortage' problem, local and transient, which in the past has found its solution in price changes which have brought supply and demand back into balance. The terms of the materials problem we face today are larger and more pervasive. (1987: 2)

'Today' for that report was over forty years ago (June 1952). Resources for the Future (RFF) was the institutional response the Paley Commission proposed to anticipate natural resource problems. It was to be an entity, detached from government, conducting research that would address the interactions the Commission had identified between people and natural resources.

About a decade later (1963) the volume *Scarcity and Growth* by Harold Barnett and Chandler Morse summarized the research undertaken at RFF on the materials problem. These authors concluded on an optimistic note, observing that:

> the resource problem is one of continual accommodation and adjustment to an ever changing economic resource quality spectrum. The physical

properties of the natural resource base impose a series of initial constraints on the growth and progress of mankind, but the resource spectrum undergoes kaleidoscopic change through time. Continual enlargement of the scope of substitutability – the result of man's technological ingenuity and organizational wisdom – offers those who are nimble a multitude of opportunities for escape. The fact of constraint does not disappear; it merely changes character. New constraints replace old, new scarcities generate new offsets. (1963: 244)

None the less, their closing chapter did identify some reasons for caution, deriving largely from what they described as indirect effects of the adjustment and technological change. In their evaluation, these effects seemed to deteriorate the quality of life. Informal evidence suggested at the time that reductions in the quality of environmental resources and other amenities were accompanying the technological responses the US economy had made to meet the need for increasing expansion in the material resources.[3]

While some of the subsequent periods of resource concern in the past thirty years have also been motivated by limitations in natural resource commodities (e.g. oil shortages during the mid-1970s), each episode has raised incrementally the profile of environmental concerns.[4] Now environmental resources are the focus of discussions about the resources problem. Today's concerns acknowledge what Barnett and Morse identified as 'nimble adjustment' but ask instead if the ways developed economies have responded are sustainable, when we consider their external effects on the environment (see Solow, 1992, as one example). One of the best examples of the difficulties in answering this query arises with ecological resources. It is for this reason (and others) that there has been increasing interest among economists in developing values for ecological resources. Such values would provide a basis of balancing the genuine material needs of the growing worldwide populations with the important need to maintain ecological resources as natural assets.

In what follows I will describe the economic framework for valuation; attempt to explain the reasons for disputes over the application of this framework to ecological resources; and describe what is known about the economic value of ecological resources.

## 12.2 THE ECONOMIC FRAMEWORK FOR VALUATION

Economic measures of the monetary value an individual places on anything (e.g. marketed commodity or non-marketed service provided by an ecosystem) result from a 'thought experiment' or abstraction on the part of the economic

[3] See Barnett and Morse (1963), pp. 254–61 for a discussion of these qualifications and Smith (1979) for a set of essays that re-examine these issues a decade and a half after the Barnett–Morse analysis was published.

[4] See Smith and Krutilla (1984) for an overview of these environmental concerns and the lack of research to address them from the vantage point of the 'end' of the oil-induced concerns about natural resource availability.

analyst. People are assumed (in the simplest economic model of individual behavior) to allocate their available income and time so as to realize the greatest level of satisfaction possible within their means. This constrained maximization process implies it is possible to describe the amount of income required, given all the external constraints affecting what is feasible for each person, for each possible level of satisfaction or utility (in the economist's vocabulary). This relationship is labeled an expenditure function.

Changes in the external conditions influencing what are feasible choices for each person will alter these income (or expenditure) requirements. One of the key supplements to this description of behavior added by environmental economists is the recognition that many dimensions of the natural environment contribute to people's satisfaction. They are often outside a person's control. None the less, when altered, as for example the severe storms in the northeast United States during the winter months of 1993 and 1994, people must adjust, changing their expenditures and time allocation decisions to do the best they can within the new conditions. With constant incomes, most are unlikely to do as well as they would have with milder winter conditions. To maintain these past levels of well-being would require more income. Economic measures attempt to capture the idea that realizing one's goals will require greater resources if the factors constraining behavior become more stringent. The thought experiment underlying expenditure functions does not imply each person's income must change to estimate the function. The implied income needs are derived using the constrained optimization framework, along with each person's observed choices.

Comparing the expenditures required for one set of environmental conditions to those of another, provided all other influences are held constant (including the level of satisfaction a person realizes), defines the monetary increment required to sustain utility. Consider offering an individual an improvement in an aspect of the environment relevant to his decisions; the expenditure difference in this case would measure the maximum amount this individual would pay to obtain it. This result follows because were it to be paid the individual's level of satisfaction would **not** change. Paying less than this maximum amount would have increased utility and more, decreased it. The same logic could be used to define monetary compensation for an undesirable change.

Three factors are important in this definition. First, it requires we accept a specific model for describing how people make choices – both in terms of their goals and what constrains their realizing them. Secondly, it acknowledges that what is important to people in life is much more than they can purchase in markets. In this expanded description of behavior, the economist can acknowledge that caring deeply about a wide range of environmental resources is an economically relevant motivation that does not require analysis outside the conventional economic paradigm. And, finally, it is precisely because life is composed of material needs, satisfied in part by allocating money to meet them, that the analyst can conceptualize people's decisions in an economic thought experiment. The tradeoff arising from a change in the environmental resource can be compared, in terms of its 'opportunity cost', with one or more other commodities also contributing to well-being, but available through markets. By expressing the results in monetary terms, the thought experiment defines the economist's concept of the willingness to pay.

This framework is not one newly discovered by environmental economists. Monetary measures of economic value were formally defined over 50 years ago by Hicks (1943). Until recently their primary application was to evaluate policies that changed the conditions of access (i.e. prices, times of availability, etc.) for marketed goods. Environmental economics changed the focus to how these measures could be adapted to define the economic values of 'things' available to people outside markets.

My selection of the term 'things' is deliberate because the transition from concept to measurement requires that the economic analyst describe how people conceptualize the commodity (or 'thing') derived from an environmental resource or ecosystem. Moreover, operational implementation of the economic framework for resources available to people outside organized markets requires detective work. When this sleuthing relies on the logic of revealed preference, it seeks to identify choices people make that can be linked to the environmental resource of interest. Methods that rely on this framework include: (a) using travel costs to recreation sites and variations in usage levels to estimate the demand for these sites; (b) exploiting spatial differences in some environmental resources and the markets for housing to relate housing prices to site-specific measures of the levels of these resources (along with the other characteristics of the site and the house); and (c) actions or expenditures consumers undertake to improve some aspect of environmental quality as it influences their other activities.[5]

In each case, the analyst must isolate the linkage and have measures of the resource's services (or quality) that are meaningful to the people making the choices these models seek to explain. While there are several important implications of this assumption, two bear directly on measures of the economic value of ecosystems. The first arises from the need to construct a linkage to some observable choice. This leads to a focus on 'use' related values. The second is more complex and arises from concerns about protecting natural ecosystems, whether tropical rain forests or marine sanctuaries, for reasons that do not have implications for current usage of the 'linked' goods. These reasons for people's concerns cannot be reflected in decisions about the related commodities. None the less, they are consistent with describing another type of service provided by ecosystems – one that can, in principle, be available to

[5] A review of each of these approaches could easily occupy several volumes. See Braden and Kolstad (1991) and Freeman (1993) for descriptions of each approach and examples of how they have been used.

As described below, there are few examples where any of these methods have been applied to evaluating changes in integrated ecosystems (aside from wetlands and specific parks or national forests).

one individual without diminishing what is available for the next person. This non-rivalry in consumption, together with an inability to exclude people from knowing that ecosystems are preserved, is one basis economists have used to argue that these types of services resemble public goods. For some people these services can be the source of the existence values Krutilla (1967) identified as missing from conventional benefit–cost criteria. These values will not be recovered in the usual revealed preference logic (see Hanemann, 1988) and require methods for eliciting directly people's choices if they are given the opportunity to influence which of these services should be provided.

Sometimes economic analyses distinguish the services assumed to underlie existence value from the role attributed to ecosystems as leading environmental indicators. That is, as indicators they may signal problems caused by current production and consumption decisions (see Randall, 1993). Discussions of the need to maintain biodiversity sometimes motivate their argument by suggesting that biodiversity serves as a gauge for the global ecosystem's resilience. Under this view, systems that have less diversity are more likely to be subject to catastrophic shocks. Thus, efforts to maintain diverse ecosystems provide society a type of insurance even if there are no directly linked private consumption goods we can attribute to particular ecosystems. Instead, people demand maintenance of diversity because they value the insurance. At a conceptual level this distinction can be important to how we characterize what a policy is providing. However, at a practical level the existence of many (sometimes incompatible) reasons to be concerned about ecosystems is a reflection of limitations in our understanding of how these natural assets in fact connect to our environment's life support services.

In many respects there are parallels between early development of economic models for describing the role of incentives to reduce effluents and current economic models of ecosystems as potentially creating a key constraint to human activity. That is, early models of residual management recognized the importance of heat and materials balances in production and consumption activities to what are feasible responses to any policy at the plant or firm level. Conventional (neoclassical) economic models did not incorporate these constraints. As a result, the responses they described to residual policy initiatives could easily be infeasible, given the physical laws governing the conversion of energy and material resources. The net result of these conflicts was an early literature suggesting economic models could not accommodate the complexities associated with real world externalities. This conclusion was not warranted. The neoclassical descriptions were not completely wrong (as some ecological economists have suggested recently). Rather they illustrate that in every field there are issues that can be addressed with general models and others requiring greater detail. These physical constraints may become especially important at particular levels of economic activity. Yet they would not be fully appreciated in conventional models relying on past operations (where behavior has taken place and the constraints were not binding).

The same observation could easily be made for the constraints associated with the regional ecosystems as natural assets contributing to economic activity and people's well-being. It is this issue that has been highlighted by a number of authors (for example, see Costanza, 1991, and Folke and Kåberger, 1992) as distinguishing the paradigms associated with conventional economics, conventional ecology and ecological economics. Fundamentally, what is at issue is whether the role of ecosystems extends beyond the status of an important **constraint** to people's activities. As Folke and Kåberger suggest, the conventional ecologist and ecological economist would argue that status as an important constraint to economic activities is not enough. In their view because people's preferences evolve in response to ecological opportunities and constraints, these preferences cannot serve the guiding role suggested by conventional economics.

Environmental economists would disagree with this perspective, acknowledging the need to supplement non-market valuation and economic analyses so that:

1. people are informed about consequences of their activities;
2. the signaling properties of markets for private goods are improved; and
3. long-term policy decisions recognize the consequences and the associated difficulties posed by inter-generational tradeoffs. None the less, within the economic perspective, decisions about the importance of resources must ultimately rely on informed choices (including revealed and stated preferences) as providing the basis for valuing market and non-market resources.

What is especially important about this approach for valuing ecological resources is the need to describe accurately the implications of these resources for people. If people's decisions are to reflect concerns about the viability of ecological resources, as assets providing services that contribute to material well-being, then they must appreciate the nature of the linkages between these assets and all the goods and services important to them. Thus, the second implication of this approach is that our lack of understanding of ecological constraints affects *both* the more private, use-related concerns and the need for better approaches to measure what people's values would be for the public good services provided by these resources.

## 12.3 FROM CONCEPT TO PRACTICE WITH ECOSYSTEM VALUES

As the previous section suggested, economic values relate to commodities – the goods and services hypothesized to contribute to individual well-being. Practical implementation of any valuation effort requires that these commodities be defined. The interrelationships between economic and ecological models depend on how these definitions enter yet another type of linkage – between economic commodities and the 'endpoints' and 'functions' described by ecological models (Russell, 1993).

For the most part economic approaches that have been argued to provide a basis for valuing ecological resources have adopted one of four commodity definitions. They focused on: one or more species, a specific land-based resource (e.g. a forest or wetlands), the perceived quality of a specific resource, or some form of 'natural insurance'. To understand why ecologists and environmental philosophers have been critical of the contribution of economic models for valuing ecosystems, we need to consider each of these types of applications, what they capture and what they miss, as well as the prospects for linking them to ecological models.

The first type of economic model seeks to estimate people's values for increments to (or avoiding reductions in) individual species. The particular strategy used depends on how the species relate to people. Thus, species often involved in consumptive uses, such as hunting or fishing, are evaluated as a yield (or as a quality dimension) of activities associated with these consumptive activities. Where the activity is non-consumptive – bird-watching, whale-watching, or simply knowledge of the existence of a species – the valuation process must establish a connection between some change resulting from a policy (e.g. a management action or investment to protect the habitat) and the outcome hypothesized to be of interest to people. For example, restrictions on the harvesting of whales will ultimately increase the population of whales. With greater abundance, threats to their extinction are reduced, and it may well be that this change increases their abundance in areas where sightings attract recreationists.

There are two aspects of these descriptions. The mechanism giving rise to the increased amount of the economic commodity rarely recognizes the other effects (or requirements to assure success) of the policies. Protecting old growth forests to preserve habitat for a bird species also affects *other* species that are maintained in these forests. Moreover, it affects the prospects for providing particular types of landscapes. The projected growth in fish available for recreation, as a result of policies to reduce commercial harvests, relies on the continued functioning of the marine environment supporting those fish populations.

Because ecologists tend to examine such issues within integrated models where multiple endpoints change, it might seem (to the external observer) that the economist is hypothesizing these other changes have no value to people. This conclusion would not be correct. Rather the economic view derives from a methodological strategy. Economists often focus on a single element (or commodity) in a partial equilibrium setting. Recognition of the need to be more general, to facilitate these other linkages, could be accomplished by changing the definition of the commodity or by adopting a framework that seeks to evaluate sets of commodities as vectors (i.e. a general equilibrium perspective). Of course, to do so requires that the proposed changes being valued authentically reflect what is feasible and that we have some basis for knowing that the people's responses are consistent with a recognition of all the elements involved in the change.[6]

Economists are increasingly calling for a type of 'consistency check' on non-market values because there are no institutions providing the equivalent of price information that allows checking our measures of economic values for environmental services. These proposals for alternative checks have suggested that efforts to value any environmental service such as maintaining a species must also evaluate the properties of the valuation measures as the 'amount' of the commodity changes. Of course, these properties would be relevant to any approach used to measure the economic values for the commodities. Concerns raised about the reliability and validity of the use of one particular method, contingent valuation, to measure non-use values has heightened interest in these properties.[7] None the less, it is important to recognize that economic theory offers rather limited guidance about what we can expect from monetary measures of value. Rational behavior, together with the other assumptions underlying our model of consumer choice, implies each individual should have a greater total monetary value for a large amount of a commodity than for a smaller amount. However, before evaluating whether a particular set of estimates meet this criterion, we must agree on a commodity definition. This definition determines how we measure 'more' of a commodity. Simple characterizations by a count of birds saved from harm (see Desvousges *et al.*, 1993) or a count of wilderness areas (Diamond and Hausman, 1993) do not necessarily correspond to the ways people think about these types of commodities.[8]

The second definition used for commodities that can be loosely related to ecosystems seems to avoid some of the need for this complete enumeration of services by identifying them as classes of natural assets. Usually they are specific areas so there is a spatial identification of the resource and a specific set of other characteristics (e.g. particular forests, fresh water marine wetlands, etc.). Economic values are derived based on observable use, or, in the case of the contingent valuation, the measures rely on description of changes to a

[6] This is a potentially difficult task as suggested by Russell (1993). The contrast between the results reported by Desvousges *et al.* (1993) and Diamond and Hausman (1993a) for specific injury descriptions associated with losses of migratory waterfowl or wilderness areas in comparison to the large scale study by Carson *et al.* (1992) indicates that to use contingent valuation for such measures will be difficult and may require extensive survey development along with in-person interviews to communicate complex, multiple endpoint commodities.

The indirect approaches may also not be superior because of correlation between the attributes (or services) that are changing. One approach that may hold promise for some types of applications uses a household production function to link the physical changes to each individual's ability to accomplish specific objectives and then relates those desired goals or outputs to a preference based decision. See Smith *et al.* (1993).

[7] For a discussion of the contrasting views on how the properties of WTP estimates can be used to evaluate valuation methods, see Hanemann (1994) and Diamond and Hausman (1993b).

[8] For a discussion of the imlications of different treatments, see Hanemann (1994) and Kopp and Smith (1993), Chapter 7.

specific resource. People's values for the natural asset depend on the services they assume it provides. Thus, we do not completely escape the need to define the commodities both as assets and their services. With these strategies, we change the perspective from valuing an asset's services to one that makes the spatial dimension (and the associated specificity in the definition of the natural asset) the primary concern. Linkages between this resource and the other dimensions of the regional ecosystem (whether positive or negative) are usually not considered. Moreover, the implications of current patterns of use for the future 'carrying capacity' of the natural asset are often implicitly assumed to have limited effects.

As a rule, assumptions about the physical carrying capacity and importance of feedback effects are simplifications used in the economic analysis to focus on what are argued (by economists) to be the important economic issues. In most cases, they will not be important to the measurement of economic value for the services of ecosystems. But they will be important to the way these values are transformed to estimate asset values. For example, following early work by Lynne *et al.* (1981), one economic approach for estimating the asset value of coastal wetlands was based on hypothesizing that they provided habitat for a wide range of fish species. Using this logic, a simple version of an economic production function was used to link measures of fish populations to the acres of coastal wetlands. Based on estimates of the consumer surplus for recreational fishing and market prices of fish, it was argued that this 'production effect' (i.e. the increase in fish stocks due to more wetlands) allowed analysts to estimate the per acre value of wetlands (i.e. the monetized value of enhanced fishing) that was argued to arise because an additional acre of wetlands was available. The asset value arises from the present (discounted) value of the increased fishing values generated each year by that acre's effects on fish stocks.

In this formulation, the effects of coastal wetlands should depend on the size of the population of fish when it is introduced, as well as any other actions influencing the marine environment that supports these fish populations. To the extent harvesting patterns in any year reduce the population below what is necessary to sustain existing fish populations, then the marginal contribution of wetlands will be different in future years than it would be if this past harvest level is lower. Of course, this same reasoning applies to other unrelated disruptions to the species' food supply that might arise as another separate consequence of the level of economic activity. In short, general equilibrium effects (in the economists' vocabulary) are not limited to market interactions. Regional ecosystems can generate close linkages between human, market based activities that would not be considered closely related from an economic perspective.

In estimating values for environmental resources, measures of these resources' services and descriptions of the 'quality' of related commodities are often used as synonyms. Marine fish species could be addressed as a species valuation task or in terms of an improvement in the 'quality' (usually measured by expected catch for a given amount of effort) of sport fishing. The issue in quality valuation that is important to my taxonomy of economic valuation strategies relates to the measure selected for quality in comparison with technical measures of the 'health' or status of the ecosystem. Perceptions of landscape amenities derived from a forested environment may well depend on both the variety of species and the age mix that is present (see Daniel and Boster, 1976; Schroeder, 1986). This could be quite different from a technical evaluation of the extent of biodiversity (see Solow *et al.*, 1993) or a technical measure of the species and age/class mixes represented in that same area. What is important in considering the compatibility of economic and scientific information is the ability to connect measures of a resource's 'services' derived from each discipline. These connections assure consistency between value measures based on perceptions and any translation of those values that are intended to represent the value for a change in the technical characteristics.

The last approach to valuing individual ecosystems treats them as indicators of the ability of the global ecosystem to continue to support mankind's activities. Implicit in this argument is the idea that once we adopt policies that avoid serious disruptions to smaller systems we develop a process that avoids large scale effects. Because we do not know whether there is such a correspondence between the local effects we observe and the larger consequences we hypothesize, a key question in valuing policies designed for local protection is whether we should attribute the value for insurance against serious aggregate impacts to the protection of local ecosystems. If the answer is yes, how much local protection is necessary to achieve a desired level of aggregate insurance (whether at a regional or global scale)?

This type of rationale parallels arguments used for avoiding continued accumulation of greenhouse gases when there is such great uncertainty about their effects on the global climate regime. It also parallels Daly's (1984) contention that conventional economic models cannot determine the appropriate aggregate scale of human activity. In his view this decision must be based on measures of the physical carrying capacity of the global ecosystem. Having outlined the general argument, it is important to recognize that all of this research remains largely at a conceptual level. Operational approaches for using it to establish the feasible scale or to determine how local protection yields global insurance have not been developed.

The first three approaches could, in principle, be linked to ecological models, at whatever scale desired. Such connections would be required to develop meaningful measures of the economic values of changes to ecosystems. Moreover, these connections are an essential first step in considering the Pearce–Turner (1990) question of whether ecosystem health, the sustainability of the interactions between economic and ecological systems, depends on the scale and the mix of mankind's economic activities in relation to the ecosystem endowment.

At present, the available applied models in ecology and economics are incompatible. As Russell (1993) suggests, the endpoints and functions of ecological models do not readily connect with the needs for economic valuation. Moreover, ecological models are not designed to accommodate the diverse set of interactions that result from people's production and consumption activities and spatially defined ecosystems. Thus, it will be difficult to use the models and results currently available to address the policy issues that motivate ecosystem valuation. The available economic estimates are also quite limited. The economic methods have tended to value one commodity at a time. The assumptions underlying these estimates suggest that they are not additive. Thus, existing economic estimates would not permit valuation of an ecosystem as the sum of the multiple endpoints currently represented in existing ecological models.

As a consequence, any comparisons of the 'valuations' developed by ecological economists and environmental economists are not likely to provide an informative basis for linking ecological models to economic values. When each group attempts to value components of an ecosystem, the starting points (economic versus ecological models) usually imply that the results will be disparate. Moreover, these differences are due both to each group's representation of the commodities, endpoints or functions associated with the ecosystems under study and to their assumptions about the relevant source of monetary values.

## 12.4 ECONOMIC VALUES OF ECOSYSTEMS: WHAT IS KNOWN?

To my knowledge, there are no measures of the economic value of full ecosystems. When economic values of these resources have been evaluated, it usually has been in terms of a narrow range of effects. Freeman (1993), for example, describes ecosystem impacts that have been subjected to economic valuation in terms of the economic productivity of the systems (i.e. agricultural productivity, forestry, and commercial fisheries), their role in supporting recreation, and with a category labeled as ecological diversity. Past economic research has not attempted to represent any of the resources supporting these impacts or effects as integrated systems. Thus, when ecosystem values are reported, they usually follow the first two of the economic commodity definitions described earlier. This is illustrated in Navrud's (1993) recent overview with his summary of the estimates of the value of maintaining habitat for particular wildlife species, protecting (or adding land to) specific natural environments (such as the Kakadu National Park in Australia) that may provide habitat as well as support a range of recreational activities, and retaining (or enhancing) the outputs attributed to wetlands.

In some cases the monetized values for services over time have been aggregated to develop an implied 'asset value' of an acre of wetlands. These capitalized values of the hypothesized stream of services from particular types of wetlands are then reported on a per acre value. Anderson and Rockel's

(1991) review of the available literature on wetland valuation is one such example. However, these computations should not be interpreted as implying that each individual acre of wetland would generate the estimated asset values. The estimates identify a wide range of services provided by a wetland with valuation estimates that are from diverse sources. No attention has been given to the likely mix of services that would be available in any specific location. Unfortunately, based on existing research alone, there is little hope for anything better than a comparably disjointed listing of services when dealing with other ecosystems. To improve matters, as Russell suggested, there must be greater opportunities for collaborative research that supports and emphasizes an integrated approach to modeling and valuation.

The debate between ecologists and economists has been misplaced. To some extent, it reflects a misunderstanding of the insights available from each discipline. Economic models do not, as Costanza (1991) seems to suggest, require limitless substitution. They do incorporate recognition of people's concern about many dimensions of the natural world. Likewise, ecological models do not require a biocentric perspective on societal objectives, with all species given equal weight in determining resource allocation decisions. Increased collaborative research can reduce such misconceptions.

However, there remains an issue that will not be resolved through enhanced communication. In judging the health of an ecosystem and its importance, whose values will determine the metric used to evaluate its status? Ecological economists contend that the objective function for analysis should be based on ecological models. Constanza (1990) described the logic underlying this conclusion noting that:

> Ecological systems are our best current models of sustainable systems. Better understanding of ecological systems and how they function and maintain themselves can yield insights into designing and managing sustainable economic systems.

Moreover, he observes that one of the conclusions that follows directly from using these systems to provide a role model for economic activities would be that sustainable economic systems should close all cycles leading to disposal of pollution. Policies should find economic uses and recycle all pollution rather than disrupt any existing ecosystem. This type of conclusion implies that the primary source of value to society is derived from maintaining functioning ecosystems in their 'natural' condition. Disruption to economic activities to avoid interfering with these ecosystems must be tolerated regardless of its costs to people. This conclusion would not be consistent with an economic evaluation of the impacts of pollution on ecosystems. Moreover, it is not simply an issue of accurately representing the activities studied by each discipline. **It is a value judgment.**

We could conceptualize the same sustainable management problem by formulating society's objective as one that minimizes disruptions to natural ecosystems subject to the constraint that each individual's level of well-being is not allowed to decline. This constraint (in a static analysis) would imply that disruptions would be permitted if there was no other way to maintain people's well-being. In this framework, ecologists would still prescribe society's objective function for sustainable use of the global ecosystem, but economic values would be recognized as the appropriate basis for making tradeoffs between people and nature. In effect economic values provide the metric for important disruptions to the health of ecosystems.

Of course, in practice, neither discipline provides *the* organizing principle for how the interactions of people and ecosystems will take place. In free societies, these judgments arise from a collection of uncoordinated activities that are constrained by the rules we collectively adopt to constrain behavior. In this setting, improving the communication between disciplines, designing research focused on improving the linkages between economic commodities and ecological endpoints and functions, and evaluating the effects of different characterizations of the way society should make these decisions, are all desirable activities. Any one of them is more likely to improve the information used in policy, and therefore ecosystem health, than further debate over whose definition of values should rule the day.

### *ACKNOWLEDGMENTS*

Partial support for this research was provided by the University of North Carolina Sea Grant Program Project No. R/MRD-21. Thanks are due to an anonymous reviewer and Richard Di Giulio for helpful comments on an earlier draft, and to Carla Skuce and Paula Rubio for preparing and editing several earlier drafts.

## REFERENCES

Anderson, R., and Rockel, M. (1991) *Economic Valuation of Wetlands*, Discussion paper No. 65, American Petroleum Institute, April 1991.

Barnett, H. J. and Morse, C. (1963) *Scarcity and Growth: the Economics of Natural Resource Availability*, The Johns Hopkins Press, Baltimore, MD.

Bell, F. (1989) *Application of Wetland Valuation Theory to Florida Fisheries*, SGR-95, Sea Grant Publication, Florida State University, Tallahassee, FL, June 1989.

Braden, J. B., and Kolstad, C. D. (eds) (1991) *Measuring the Demand for Environmental Quality*, North Holland, Amsterdam.

Carson, R. T., Mitchell, R. C., Hanemann, W. M. *et al.* (1992) *A Contingent Valuation Study of Lost Passive Use Values Resulting from the Exxon Valdez Oil Spill*, Report to Attorney General of the State of Alaska, Natural Resource Damage Assessment, Inc., La Jolla, CA.

Conservation Foundation Ecosystem Valuation Forum (1992) *Issues in Ecosystem Valuation: Improving Information for Decision-Making*, Phase I, Report to the US Environmental Protection Agency.

Costanza, R. (1990) Ecological economics as a framework for developing sustainable national policies, in *Towards an Ecologically Sustainable Economy* (eds B. Aniansson and S. Sveden), The Swedish Council for Planning and Coordination of Research, Stockholm.

Costanza, R. (ed.) (1991) *Ecological Economics: the Science and Management of Sustainability*, Columbia University Press, New York.

Costanza, R., Farber, S. C. and Maxwell, J. (1989) Valuation and management of wetland ecosystems. *Ecological Economics*, **1**, 335–61.

Daniel, T. and Boster, R. S. (1976) *Measuring Landscape Esthetics: the Scenic Beauty Estimation Method*, USDA Forest Service Research Paper RM-167, Rocky Mountain Forest and Range Experiment Station, Fort Collins, CO.

Daly, H. E. (1984) Alternative strategies for integrating economics and ecology, in *Integration of Economy and Ecology: an Outlook for the Eighties* (ed. A. M. Jansson), Askö Laboratory, University of Stockholm.

Desvousges, W. H., Johnson, F. R., Dunford, R. W. *et al.* (1993) Measuring natural resource damages with contingent valuation: tests of validity and reliability, in *Contingent Valuation: a Critical Assessment* (ed. J. A. Hausman), North Holland, Amsterdam.

Diamond, P. A. and Hausman, J. A. (1993a) On contingent valuation measurement of non-use values, in *Contingent Valuation: a Critical Assessment* (ed. J. A. Hausman), North Holland, Amsterdam.

Diamond, P. A. and Hausman, J. A. (1993b) Contingent valuation: is some number better than no number? Unpublished paper, Department of Economics, MIT, Boston, MA.

Folke, C., and Kåberger, T. (1992) Recent trends in linking the natural environment and the economy, in *Linking the Natural Environment and the Economy: Essays from the Eco-Eco Group,* (eds C. Folke and T. Kåberger), 2nd edn Kluwer Academic Publishers, Norwell, MA.

Freeman, A. M., III (1993) *The Measurement of Environmental and Resource Values*, Resources for the Future, Washington, DC.

Hanemann, W. M. (1988) Three approaches to defining 'existence' or 'non-use' value under certainty. Working paper, Department of Agricultural and Resource Economics, University of California, Berkeley, CA.

Hanemann, W. M. (1994) Contingent valuation and economics. Unpublished paper, Department of Agricultural and Resource Economics, University of California, Berkeley, CA.

Hicks, J. R. (1943) The four consumer surpluses. *Review of Economic Studies*, **11**(1), 31–41.

Kopp, R. J. (1993) Environmental economics: not dead but thriving. *Resources* **111**, 7–12.

Kopp, R. J. and Smith, V. K. (eds) (1993) *Valuing Natural Assets: the Economics of Natural Resource Damage Assessments*, Resources for the Future, Washington, DC.

Krutilla, J. V. (1967) Conservation reconsidered. *American Economic Review*, **57**, 777–86.

Lynne, G. D., Conroy, P. and Pochasta, F. (1981) Economic valuation of marsh areas to marine production processes. *Journal of Environmental Economics and Management*, **8**, 175–86.

Navrud, S. (1993) Economic value of biological diversity in Norway. *Scandinavian Forest Economics*, **43**, 74–97.

Pearce, D. W. and Turner, R. K. (1990) *Economics of Natural Resources and the Environment,* Harvester/Wheatsheaf, Hertfordshire, UK.

President's Materials Policy Commission (1987) Report: *Resources to Freedom*, Summary of Volume I, (GPO, Washington, DC, 1952), reprinted by Resources for the Future, 1987.

Randall, A. (1993) Thinking about the value of biodiversity. *Scandinavian Forest Economics*, **43,** 4–17.

Russell, C. S. (1993) Old lessons and new contexts in economic-ecological modeling. Paper presented at Symposium on Integrating Economic and Ecological Indicators, Resource Policy Consortium, World Bank, 17–18 May 1993.

Sagoff, M. (1993) Environmental economics: an epitaph. *Resources*, **111,** 2–6.

Schroeder, H. W. (1986) Estimating park tree densities to maximize landscape esthetics. *Journal of Environmental Management*, **23,** 325–33.

Smith, V. K. (ed.) (1979) *Scarcity and Growth Reconsidered*, The Johns Hopkins Press, Baltimore, MD.

Smith, V. K. (1980) The evaluation of natural resource adequacy: elusive quest or frontier of economic analysis? *Land Economics*, **56**(3), 257–98.

Smith, V. K. and Krutilla, J. V. (1984) Economic growth, resource availability and environmental quality. *American Economic Review*, **74,** 226–30.

Smith, V. K., Liu, J. L. and Palmquist, R. B. (1993) Marine pollution and sport fishing quality: using poisson models as household production functions. *Economics Letters*, **42,** 111–16.

Solow, A., Polasky, S. and Broadus, J. (1993) On the measurement of biological diversity. *Journal of Environmental Economics and Management*, **24,** 60–8.

Solow, R. (1992) An almost practical step toward sustainability. An invited lecture on the occasion of the 40th Anniversary of Resources for the Future, 8 October 1992.

# 13 *Perceptions of risk to humans and to nature: a research plan*

PAUL SLOVIC, TIMOTHY MCDANIELS AND
LAWRENCE J. AXELROD

## 13.1 INTRODUCTION

Managing risk to human health and safety has, over the past two decades, become a dominant theme in government policy, public debate, media attention and academic research. A striking feature of this growth has been the increased role of social scientists (working in perception, judgment and decision-making) in debates that were initially characterized as completely based in science and technology. Examples include applications of risk-perception research (Slovic, 1987), development of normative theories regarding equity in public risk (Keeney, 1980), emergence of risk communication as a field of endeavor and research (National Research Council, 1989), characterization of 'mental models' of how lay-people and experts think about health risk decisions (Bostrom *et al.*, 1992), preparation of environmental impact statements (Gregory *et al.*, 1992), and integration of perception into the evaluation of risk-management options (McDaniels *et al.*, 1992). The application of social and decision sciences in problems of risk management has thus been substantial.

In recent years, increasing attention has been directed to risk management questions of another kind. Ecological risks (threats to the health and productivity of ecosystems and species) have arisen as a topic of great public concern, in parallel with heightened attention to sustainability and concern over environmental degradation. Examples of ecological risks range from specific threats to localized ecosystems due to, say, toxic effluents, to threats to global ecosystems from climate change. While the risk management community has recently recognized the increasing need for serious research on ecological risk

*Interconnections Between Human and Ecosystem Health.* Edited by Richard T. Di Giulio and Emily Monosson. Published in 1996 by Chapman & Hall, London. ISBN 0 412 62400 1.

management (Travis and Morris, 1992; Lackey, 1993; Suter, 1993), much of the work to date has been undertaken by physical and biological scientists. Relatively little effort has been devoted to decision science activities that tackle basic questions regarding human perception, mental characterization, value assessment, or decision-making structures for ecological risks.

This chapter explores one basic aspect pertaining to ecological risk management: the need for research designed to characterize ecological risk perception. The research activities discussed in this paper would build on the psychometric risk perception paradigm developed to characterize human health risk (Slovic, 1987, 1992), using its basic concepts and methods to describe how people perceive ecological risks. This work represents exploratory research designed to identify and clarify what people mean when they say something is 'risky to the environment'. It would contribute to a better understanding of what has been termed the 'social construct' of ecological risk. This social construct, and not objective probabilities of harm, will ultimately play the predominant role in shaping ecological risk management (Stehr and von Storch, 1994).

Because this research builds on the psychometric paradigm for human health risk perception, a brief explanation of the paradigm's underlying concepts and methods will be helpful. The psychometric approach to risk perception is a descriptive method for determining the underlying characteristics that influence people's perceptions of whether a given technology is risky to human health. The underlying conceptual viewpoint is that 'risk' is inherently multidimensional, with many characteristics other than just probability of harm, or, say, voluntariness of activities (Starr, 1969) affecting what people view as risky. This approach to characterizing risk perception has been applied to a wide range of technologies, and used in many cross-cultural comparisons to provide a detailed representation of health risk perception (Slovic, 1992).

The basic steps of the psychometric approach include:

1. Developing a list of 'hazard items' or risky activities that span a broad domain of important hazards.
2. Developing a number of psychometric scales that represent dimensions (characteristics) of risks that are important in shaping human perception of, and response to, different hazards.
3. Asking people to evaluate the list of items on each of the scales.
4. Using multivariate statistical methods such as factor analysis, multidimensional scaling and structural equations modeling to identify and interpret a set of underlying factors that capture the variation in the individual and group responses.

Figure 13.1 is a well-known representation of the results of several psychometric analyses of the perceived characteristics of various technological hazards. It shows that two underlying factors, which have been termed 'dread'

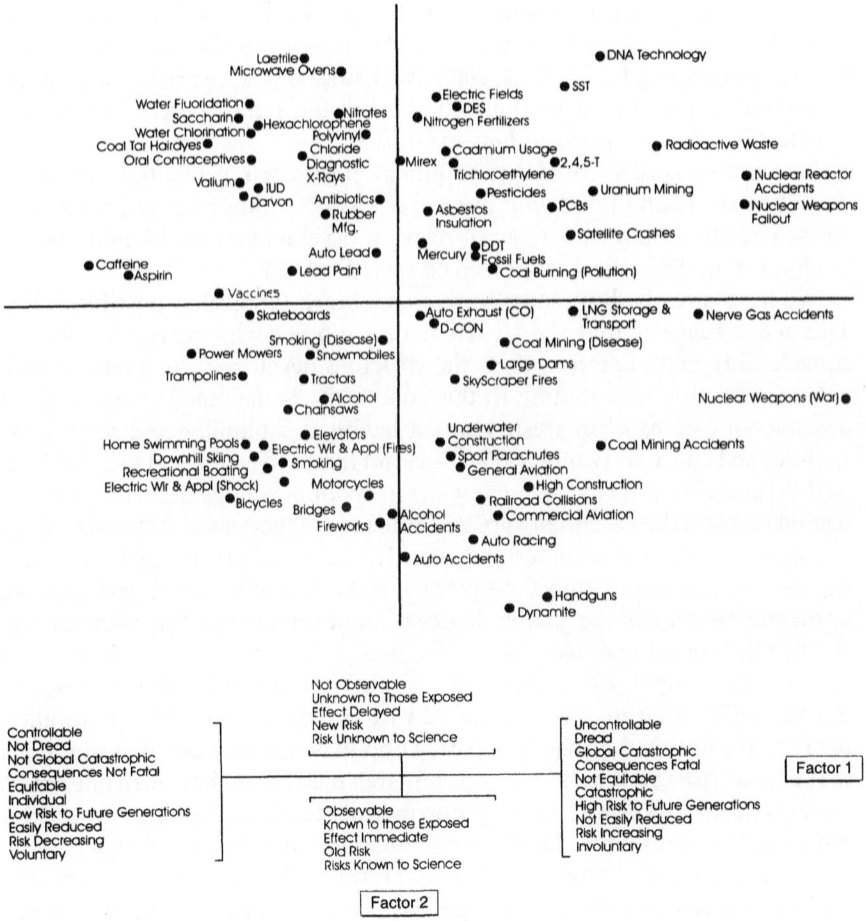


**Figure 13.1** Location of 81 hazards on Factors 1 and 2 derived from the interrelationships among 15 risk characteristics. Each factor is made up of a combination of the characteristics, as shown in the lower diagram. (Reproduced with permission from Slovic, 1987.)

and 'knowledge', are useful in creating a two-dimensional risk perception map. Details of the analytical steps in the psychometric approach are found in Slovic *et al.*, (1980) and Slovic (1992).

Numerous studies carried out within the psychometric paradigm have shown that perceived risk is quantifiable and predictable (Slovic, 1987). Psychometric techniques seem well suited for identifying similarities and differences among groups with regard to risk perceptions and attitudes. They have also shown that the concept 'risk' means different things to different people. When experts judge risk, their responses correlate highly with technical estimates of annual fatalities. Lay people can assess annual fatalities if they are asked to (and produce estimates somewhat like the technical estimates). However, their judgments on risk are related more to the characteristics of hazards underlying Figure 13.1, such as whether or not the risk is voluntary, controllable, inequitable, or dreaded. As a result, public perceptions of risk tend to differ from experts' judgments of risk.

Research is needed to adapt the human health risk perception paradigm described above and build on it in several ways. Perhaps most important will be the need to develop a new set of relevant psychometric scales that capture the important dimensions of perceived ecological risk.

We have already done some preliminary work to develop possible scales. This work suggests that the structure of ecological risk perception will be considerably more complex than the structure developed for human health risks. One factor contributing to this complexity is the much wider range of possible end states of interest. Human health is a familiar and reasonably well-defined concept. Ecological health (and threat to ecological health) is less well-defined and will have a much wider array of meanings (Lackey, 1993). A second factor is the potentially greater influence of the personal characteristics of subjects, such as their underlying worldviews, value orientations, and prior experience with nature and the potential risks. These factors could be more influential because of the greater degree of subjectivity regarding what ecological health means to people.

Another source of complexity is the multidimensional notion of 'responsibility' that may be associated with ecological risks. 'Responsibility' refers to the notion that one group may obtain benefits by creating situations that impose risks on others. Previous research has shown that when one group imposes health risks on a second group in order to obtain benefits, the second group will likely find that imposed risk unacceptable (Fischhoff *et al.*, 1981). For ecological risks, responsibility could refer to risks imposed on a specific group of humans, on non-human species, or on future generations of humans and non-humans. How people perceive their responsibilities to various groups of 'others' in ecological risk contexts, and how these perceptions shape behavior, is the essence of the social dilemmas, or 'commons' conflicts, created by open-access resources (Hardin, 1968; Dawes, 1980).

Still other sources of increased complexity will be the greater variation in the physical scales of ecological risk, which could range from affecting a few square meters of plants to global ecological change, and the likely concern with impacts on ecological systems and whole species, rather than effects on individuals or groups.

All these points suggest that care and creativity will be required to devise sets of scales that can meaningfully capture the range of possible dimensions of perceived ecological risk. In addition, a careful research plan is needed to ensure that information about personal characteristics of subjects (e.g. worldviews, value orientations and other information) is obtained to provide a more complete picture of the factors shaping ecological risk perception.

## 13.2 APPROACH

There is no simple or established formula for designing a framework for characterizing ecological risk perception. Such characterization requires a blend of educated intuition (to select hazards and to design scales with which to characterize these hazards) and sophisticated analysis of the resulting data matrix. The factor space shown in Figure 13.1 is a pictorial representation of the relationships among more than 40 000 judgments. The massive reduction of the data matrix leading to this two-dimensional plot is both a strength and a weakness: the simple visual representation conveys a great deal of meaning, but the representation is clearly a superficial characterization of the complex world of perceived risk.

Risk-perception surveys have been continually revised over the years to create improved representations. Later studies have involved new hazard domains and new populations of respondents from around the world. Factor analysis and standard descriptive statistics have been augmented by use of powerful multivariate techniques such as three-dimensional factor analysis and multidimensional scaling techniques (e.g. INDSCAL) that allow individual differences in perceptual representations to emerge (see, for example, Arabie and Maschmeyer, 1988). These exploratory forms of analysis have been supplemented by use of confirmatory analyses such as LISREL (Fornell, 1982), designed to test specific theoretical models (Flynn *et al.*, 1992). Thus the program on perception of risk to humans began with simple exploratory studies and has evolved toward the creation and testing of specific theories.

We believe that work on ecological risk perception will need to follow the same progression, beginning with the explorations we plan to carry out in the near future. Efforts to create scales and hazard items should begin with focus groups, asking small groups of local citizens to participate in semi-structured discussions about what 'ecological risk' means to them. These processes can be supplemented with an examination of the small but growing technical literature on ecological risk assessment (e.g. Lackey, 1993; Suter, 1993); much health risk perception initially drew on the technical analyses of human health and safety risks by Lowrance (1976) and Starr (1969). Literature on environmental ethics and values (e.g. Rolston, 1981; Armstrong and Botzler, 1993), the literature on environmental psychology (e.g. Kaplan and Kaplan, 1982), and studies of attitudes toward nature (for example, Kellert's studies of perception of animals; Kellert, 1984) can also contribute to devising scales. Finally, the burgeoning field of nature writing (Slovic and Dixon, 1993) can be examined for insights that might be contained in the observations of some of our most eloquent writers and thinkers.

## 13.3 HYPOTHESES

Several preliminary hypotheses can be proposed as a guide to these research activities. One hypothesis is that at least some of the underlying factors that have been shown to be important in risk perception of human health will also be important in ecological risks. These underlying factors might include the degree of dread, degree of knowledge and degree of control that individuals associate with the various hazards. On the other hand, new factors may arise as well. One might be the scale of the impact. In general terms, one could expect that impacts that are potentially global in nature will be viewed as more risky than those that are likely to be localized.

A second hypothesis is that people will generally view activities that threaten an entire species or a large ecosystem as highly risky, but not view threats to individual animals or plants as particularly risky. This is in contrast to human health risks, in which protection of individuals is extremely important. A third hypothesis is that there will be substantial agreement among experts and lay people regarding what is ecologically risky and what is not. This would contrast with the findings in the human health risk profession, where experts tend to rate riskiness on the basis of statistical estimates, while lay people judge riskiness on the basis of factors such as the degree of dread and the degree of control over the risk. We also expect that there will be greater perceived ecological risks associated with human activities than with natural disasters, and greater perceived risks from highly publicized global threats (e.g. climate change) than from the activities that directly contribute to those threats (e.g. use of coal-fired power plants). Finally, we expect that the complexity of ecological risk will cause worldviews and trust in responsible authorities and institutions to be even more important as 'orienting dispositions' in this domain than they are in the realm of human health risk perception.

## 13.4 SPECIFIC RESEARCH STAGES

Based on the information gleaned from the sources described above, psychometric studies will involve several stages, as outlined below.

**Stage 1 Compile a list of 'hazards to nature'**. We expect this list to have several major categories such as:

A. Human activities: which might include such items as those shown in Table 13.1.
B. Natural phenomena or events: which might include such items as floods, hurricanes, asteroids, volcanoes, cosmic radiation, lightning, earthquakes and UV radiation.
C. Specific consequences of A or B above: which might include human population growth, increased greenhouse gases, global temperature increase, sea level rise, and introduction of various kinds of exotic organisms into an ecosystem.

**Table 13.1** Human activities that are potential 'hazards to nature'

| | |
|---|---|
| • Drift net fishing | • Disposing of treated sewage in lakes and rivers |
| • Clear cut logging | • Various kinds of genetic engineering |
| • Selecting logging | • Garbage incinerators |
| • Sport fishing | • Microwave radiation for communications |
| • Commercial troll fisheries | • Naturally occurring radiation |
| • Strip mining | • Irrigated agriculture |
| • Underground mining | • Domestic use of water from rivers |
| • Wilderness hiking | • Pumping ground water for domestic use |
| • Helicopter skiing | • Cloud seeding |
| • Wildlife viewing | • Building roads for forest harvest |
| • Pesticides in commercial agriculture | • Building roads for fire access |
| • Pesticides in home gardens | • Fighting forest fires |
| • Toxic waste dumps | • Mass tourism |
| • Hazardous waste disposal facilities | • Eco-tourism |
| • Nuclear waste repositories | • Paper recycling |
| • Municipal landfills | • Mineral exploration (or acid mine drainage) |
| • Chlorofluorocarbons in consumer products | • Slash and burn agriculture in tropical forests |
| • Coal-fired power plants | • Paving roads |
| • Wood burning fireplaces | • Urban runoff |
| • Using natural substances to combat garden pests | • Excavating holes for buildings |
| • Nuclear power plants | • Naturally occurring poisons |
| • Collecting clams | • Mountain bikes |
| • Collecting insects | • Reintroduction of wolves in the western states |
| • Hunting large mammals | • Wetlands reclamation |
| • Hunting birds | • Canoeing |
| • Hunting for ceremonial uses | • Power boats for recreation |
| • Downhill skiing | • Oil transport by tankers |
| • Cross-country skiing | • Chemical plants |
| • Driving cars | • Pulp mills |
| • Driving a heavily polluting car that needs a tune-up | • Petroleum refineries |
| • Harvesting mushrooms | • Airplanes |
| • Harvesting wild plants | • Airports |
| • Collecting wildflowers | • Dry-cleaning plants |
| • Radon gas in caves | • Power lines |
| • Methane in caves | • Pouring used motor oil on vacant land |
| • Oil exploration | • Pouring used motor oil in storm sewers |
| • Large dams on rivers | • Nuclear bombs |
| • Burning natural gas in homes | • Conventional warfare |
| • Disposing of untreated sewage in the ocean | • Urban water infrastructure |
| • Disposing of treated sewage in the ocean | |

**Stage 2 Create scales on which people can characterize the hazard items identified in Stage 1.** For example, one possible scale might be:

'To what extent does this (activity/phenomenon) endanger an entire species?' (Circle one number.)

| 1 | 2 | 3 | 4 | 5 | 6 | 7 |
|---|---|---|---|---|---|---|
| **no threat** to an entire species | | | | | **great threat** to an entire species | |

Each item would be rated on this scale.
Examples of other scales might include:

1. spatial extent of the hazard (very localized to global);
2. degree of threat to non-human mammals;
3. degree of threat to insects and/or microorganisms;
4. degree of threat to plants;
5. degree of threat to non-living nature (e.g. rocks, mountains);
6. degree of threat to human values: separate scales for
   (a) human health and safety
   (b) human economic well-being
   (c) aesthetic values
   (d) spiritual values;
7. degree of benefit to nature or to human values such as 2–5 and 6(a–d) above;
8. need for human intervention or regulation to reduce or mitigate risk from the activity or phenomenon;
9. ability of humans to reduce the major ecological risks associated with this hazard, and many other scales.

**Stage 3 Generate other survey questions.** Human health risk perceptions have been found to be correlated with a wide variety of personal and demographic variables such as gender, age, education, values and worldviews. It would be useful to include questions in studies that would allow one to determine whether and how ecological risk perceptions are related to these variables.

Special attention should be given to worldviews, which can be seen as general philosophies or dispositions that enable people to simplify the world and thus manage its complexities (see, for example, Dake, 1991). Worldviews are expressed in terms of general attitudinal statements that respondents are asked to agree or disagree with. Some of the worldviews found relevant to human health risk perception (along with representative statements) are as follows.

- Desire for hierarchical social relationships (e.g. 'Decisions about risks to the environment should be left to the experts').
- Individualism (e.g. 'In a fair system, people with more ability should earn more').

- Egalitarianism ('If people in this country were treated more equally, we would have fewer problems').
- Technological enthusiasm ('A high technology society is important for improving our health, our environment, and our social well-being').

Other worldviews include the clash between:

- cornucopian views (Earth is robust and rich in resources), and
- catastrophist views (Earth is fragile and its resources are limited; see, for example, Cotgrove, 1982).

Nature writers and philosophers have discussed other fundamental views that can be translated into attitude statements. For example:

- human beings are part of nature, *vs.*
- human beings are separate from and above the rest of nature.

**Stage 4 Conduct major studies.** Once the questionnaire items and scales have been selected and refined by pilot testing, the survey instrument would be ready for administration to diverse groups such as:

1. citizens;
2. biologists and ecologists;
3. persons in resource-dependent and resource-independent communities;
4. workers in state departments of environmental quality or conservation.

## 13.5 POTENTIAL BENEFITS

Judging from the analogy of human health risk perception research, the potential contributions of ecological risk perception research could be substantial, multifaceted, and enduring. The benefits of health risk perception work have largely come in three areas: behavioral insights about the characteristics of technologies that lead to heightened perceived risk from the viewpoints of various groups; descriptive insights about risk-management controversies; and prescriptive insights about preferences, risk communication practice, stigma, and the 'signal' potential of technological mishaps (Slovic, 1992).

Analogous insights could also be expected to stem from ecological risk perception research. The ecological risk perception results would provide a major step toward characterization of the underlying factors that shape perceptions of environmental risk in various groups. These characterizations will in turn lead to descriptive insights that can help diagnose environmental management conflicts and provide explanations of why controversies (largely stemming from differences in perceptions among groups) arise in some situations and not in others.

Finally, learning about ecological risk perception could provide a wide array of prescriptive insights. These insights could lead to improved communication programs for understanding ecological risks, help in predicting events that are viewed as signals of growing ecological harm, help in building models of preferences for risk changes, and help in examining the characteristics of ecological risks for which subjects indicate greater willingness to pay for risk reduction and greater willingness to cooperate with others in resolving resource conflicts. By learning more about how individuals think about ecological risks, we could expect to eventually manage these risks more effectively.

*ACKNOWLEDGMENTS*

Work on this chapter was supported by a Government of Canada Tri-Council Ecoresearch grant to Westwater Research Centre at the University of British Columbia and by an assistance agreement R-822464-01 from the US Environmental Protection Agency to Decision Research.*

## REFERENCES

Arabie, P. and Maschmeyer, C. (1988) Some current models for the perception and judgment of risk. *Organizational Behavior and Human Decision Processes*, **41,** 300–29.

Armstrong, S. J. and Botzler, R. G. (1993) *Environmental Ethics: Divergence and Convergence*, McGraw-Hill, New York.

Bostrom, A., Fischhoff, B. and Morgan M. G. (1992) Characterizing mental models of hazardous processes: a methodology and an application to radon. *Journal of Social Issues*, **48,** 85–100.

Cotgrove, S. (1982) *Catastrophe or Cornucopia: the Environment, Politics and the Future*, Wiley, New York.

Dake, K. (1991) Orienting dispositions in the perception of risk: an analysis of contemporary worldviews and cultural biases. *Journal of Cross-Cultural Psychology*, **22,** 61–82.

Dawes, R. M. (1980) Social dilemmas. *Annual Review of Psychology*, **31,** 169–93.

Fischhoff, B., Lichtenstein, S., Slovic, P. *et al.* (1981) *Acceptable Risk*. Cambridge University Press, New York.

Fischhoff, B., Watson, S. and Hope, C. (1984) Defining risk. *Policy Sciences*, **17,** 123–39.

Flynn, J., Burns, W., Mertz, C. K. and Slovic, P. (1992) Trust as a determinant of opposition to a high-level radioactive waste repository: analysis of a structural model. *Risk Analysis*, **12,** 417–30.

Fornell, C. (1982) *A Second Generation of Multivariate Methods*, Praeger, New York.

Gregory, R., Keeney, R. and von Winterfeldt, D. (1992) Adapting the environmental impact statement process to inform decision makers. *Journal of Policy Analysis and Management*, **1,** 58–75.

Hardin, G. (1968) The tragedy of the commons. *Science*, **162,** 1243–8.

Kaplan, S., and Kaplan, R. (1982) *Cognition and Environment: Functioning in an Uncertain World*, Praeger, New York.

Keeney, R. L. (1980) Equity and public risk. *Operations Research*, **28,** 527–34.

* This chapter has not been subjected to the EPA's peer and administrative review and therefore may not necessarily reflect the views of the Agency and no official endorsement should be inferred.

Kellert, S. R. (1984) Urban American perceptions of animals and the natural environment. *Urban Ecology*, **8,** 209–28.

Lackey, R. T. (1993) Ecological Risk Assessment. Paper presented at Symposium on Critical Issues in Risk Assessment, Tulane, New Orleans.

Lowrance, W. (1976) *Of Acceptable Risk*. Kaufmann, Los Altos, CA.

McDaniels, T., Kamlet, M. and Fischer, G. (1992) Risk perception and the value of safety. *Risk Analysis*, **12,** 495–503.

Merchant, C. (1992) *Radical Ecology*, Routledge, New York.

National Research Council (1989) *Improving Risk Communication*, National Academy Press, Washington, DC.

Rolston, H. (1981) Values in nature. *Environmental Ethics*, **3,** 113–28.

Slovic, P. (1987) Perception of risk. *Science*, **236,** 280–5.

Slovic, P. (1992) Perception of risk: reflections on the psychometric paradigm, in *Social Theories of Risk* (eds S. Krimsky and D. Golding), Praeger, New York, pp. 117–52.

Slovic, P., Fischhoff, B. and Lichtenstein, S. (1980) Facts and fears: understanding perceived risk, in *Societal Risk Assessment: How Safe is Safe Enough?* (eds R. Schwing and W. A. Albers), Plenum, New York, pp.181–214.

Slovic, S. H. and Dixon, T. F. (1993) *Being in the World: an Environmental Reader for Writers*, Macmillan, New York.

Starr, C. (1969) Social benefit versus technological risk. *Science*, **165,** 1232–8.

Stehr, N. and von Storch, H. (1994) *The Social Construct of Climate and Climate Change* (Report no. 137). Max-Planck-Institute für Meteorologie, Hamburg.

Suter, G. W. I. (1993) *Ecological Risk Assessment*, Lewis Publishing, Boca Raton, FL.

Travis, C. C. and Morris, J. M. (1992) The emergence of ecological risk assessment. *Risk Analysis*, **12,** 167–8.

# 14 *Perceptions of ecosystem health, stress and human well-being*

MARY K. O'KEEFFE AND ANDREW BAUM

In examining the interconnections between human and ecosystem health, careful consideration of the direct physiological consequences of toxic exposure is but one of many critical elements to be considered. Scientists must also evaluate the direct psychological and indirect physiological and psychological consequences resulting from the perception that one has been exposed to a toxic agent if we are to accurately model and predict health outcomes. In addition, associations among nervous system, endocrine system and immune system functioning make it likely that the perceptions and toxins interact in complex ways to determine the ultimate effects of exposure to environmental hazards.

## 14.1 PERCEPTIONS OF ECOSYSTEM HEALTH

Concerns about ecosystem health include fears about a wide array of environmental threats such as ozone depletion, radon gas, nuclear contaminants and waste, electromagnetic radiation and a host of contaminants in the soil, water and air. Response to such pollutants and environmental damage is varied. Some people react apathetically, ignoring both the threats posed and readily available solutions to these environmental problems. Others may react in a purposive fashion, taking direct steps to cope with and/or alleviate the problem. However, fears and worry about environmental hazards can become exaggerated and may be associated with confusion and perceptions of helplessness. It has been argued that these threats have contributed to an overall 'epidemic of apprehension' in American society, where people see threats lurking in every corner, feel powerless to modify their relationship with these various threats, and respond with either chronic anxiety about their health or

*Interconnections Between Human and Ecosystem Health.* Edited by Richard T. Di Giulio and Emily Monosson. Published in 1996 by Chapman & Hall, London. ISBN 0 412 62400 1.

with self- protective denial, either of which can lead them to ignore even the most basic health messages (Becker, 1993). Our research has indicated that perceived exposure to environmental hazards is also associated with psychophysiological consequences indicative of chronic stress (e.g. Baum *et al.*, 1983b; Baum, 1990; O'Keeffe and Baum, 1990; Baum and Fleming, 1993).

Most of us are worried about the current and future state of the environment in which we live, with public opinion polls showing that issues such as air or water pollution, contamination of soil and ground water, and hazardous waste storage and disposal are often listed as major sources of alarm and concern (Kohut and Shriver, 1989; Tyson, 1992). Our worries about the environment are not without basis, and many communities may eventually be threatened with potential exposure to some type of environmental hazard. Well over 30 000 hazardous waste areas are now being considered for Superfund sites, and it has been suggested that approximately 75 000 sites will eventually make their way to the Superfund list (Abelson, 1992). However, perceptions of environmental hazards, no matter how clear or stable, are not good markers of the impact of these hazards on behavior, mood or health. Rather they are mediators, influencing the extent to which hazards are seen as threatening or dangerous and the extent to which stress is experienced. Coping and vulnerability may also be affected by such perceptions. This makes measurement of these phenomena particularly difficult.

## 14.2 ASSESSMENT OF EFFECTS OF PERCEIVED HAZARDS ON WELL-BEING

When conducting an assessment of the impact of an environmental threat on a given population, it is necessary whenever possible to assess the direct effects that the particular toxin may have on an exposed organism. Epidemiological research and laboratory experiments with non-human subjects can assist with examination of the direct carcinogenic, mutagenic and/or teratogenic effects of exposure. However, it has been argued that this type of quantitative risk assessment is a non-exact science, given the frequent absence of accurate information on extent of exposure and/or the toxicology of the specific agents involved (Wandersman and Hallman, 1993).

Accurate impact assessments must also recognize that stress, depression, anger, loss of control and other psychological sequelae of the perception that one has been exposed to a toxin may have long-term health consequences. Our studies of residents living within 5 miles of the damaged nuclear reactor at Three Mile Island have provided evidence of chronic stress characterized by higher blood pressure, stress-related hormone levels and small differences in immune system status relative to unexposed control subject up to six years after the accident (e.g. Baum *et al.* 1983b; McKinnon *et al.* 1989; Baum, 1990). We have found similar indications of stress in populations living near other types of toxic hazards (O'Keeffe *et al.*, 1989; Baum *et al.*, 1992; Baum and Fleming,

1993). In addition, stress-linked cancers (i.e. cancers thought to be independent of presumed radiation exposure) following the TMI nuclear accident have been reported, as have health problems in children whose mothers were extremely disturbed about being pregnant at the time of the accident (Hatch *et al.*, 1991; Houts *et al.*, 1991).

Before reviewing our work on chronic stress as a consequence of perceived exposure to environmental threats, it is important to note that recognition of the psychophysiological consequences of this stress may complicate assessment of the direct toxic effects of specific hazards and lead to increased misunderstanding, public skepticism and anger. Illustrative of this was a 1991 *Washington Post* article titled 'UN blames stress, not radiation, for Chernobyl illnesses' (Wise, 1991). The article reported that the International Atomic Energy Agency (IAEA) 'acknowledged that stress-related illnesses were caused by a lack of public information about the disaster and the mass evacuations that followed, but termed these "wholly disproportionate to the significance of radioactive contamination" ' (Wise, 1991: A24). It also noted that environmental groups such as Greenpeace called the IAEA report a 'whitewash, tailored to serve the interests of the nuclear industry' (Wise, 1991: A24). Appreciation of the potential consequences of chronic stress following perceived exposure may be presented in a manner that appears to compromise the legitimacy of the complaints of potential victims and the general public often is left to interpret mixed messages presented by the media.

Other instances of this kind of complication can be found in the controversies about Persian Gulf Syndrome (malaise associated with serving in the Gulf War, etiology currently undetermined) or about the long-term effects of herbicides containing dioxin, most notoriously Agent Orange. Many believe that Agent Orange exposures were responsible for some of the long-term psychological and physical problems experienced by some Vietnam veterans and this issue continues to provoke passionate responses from veterans' organizations and federal officials (McPherson, 1984; Stone, 1992). Dioxin is indisputably a potent toxin, yet reports of deleterious effects on humans are often debated due to the difficulty in quantifying exposure (Poland and Glover, 1973; Poland and Kimbrough, 1984; US Committee on Government Operations, 1993; US Committee on Veterans' Affairs, 1994). Psychological and medical problems in Vietnam veterans have been linked to subjective but not objective assessments of exposure to Agent Orange (Korgeski and Leon, 1983). Based on assessments of only 100 Vietnam veterans, this study concluded that 'belief that one was exposed to herbicides, regardless of whether one actually was exposed, is associated with reports of a wide variety of vague psychiatric and psychosomatic symptoms' (Korgeski and Leon, 1983: 1448). However, like the hundreds of studies that have been conducted to assess the legitimacy of hypothesized links between Agent Orange exposure and health problems among veterans, results are inconclusive due not only to problems in objectively quantifying exposure but also because it is often difficult to isolate the physiological effects of exposure from the psychophysiological consequences of believing that one has been exposed to a toxin.

## 14.3 CHRONIC STRESS AS A CONSEQUENCE OF PERCEIVED TOXIC EXPOSURE

Exposure to human-made hazards (e.g. leaking toxic landfills, damaged nuclear reactors and dangerous chemical spills) is associated with acute psychological disturbances characteristic of exposure to natural disasters (e.g. tornadoes, floods, earthquakes) such as fear, anxiety, denial and depression (Quarantelli, 1985; Sowder, 1985; Baum *et al.*, 1983b; Baum, 1990; Bromet *et al.*, 1990). In addition, a range of psychological responses are associated with these catastrophes.

1. Perceived loss of control over the environment and resulting helplessness.
2. Outwardly directed anger and blame.
3. Uncertainty about future health threats and fear of cancer.
4. Continuing intrusive images about the exposure.

These sequelae contribute to the chronic psychophysiological symptoms seen following exposure to many human-made hazards (Davidson *et al.*, 1982; Baum *et al.*, 1983a; Baum and Fleming, 1993; Baum, 1990; Wandersman and Hallman, 1993).

Much of what we know about long-term stress responding following exposure to a human-made hazard comes from a series of studies of residents living near the damaged nuclear reactor at Three Mile Island (TMI) conducted by Baum and colleagues throughout the 1980s (e.g. Davidson *et al.*, 1982; Baum *et al.*, 1983b; Gatchel *et al.*, 1985; Davidson and Baum, 1986; McKinnon *et al.*, 1989). This work grew out of an interest in extending disaster research to include more careful consideration of the psychosocial and psychophysiological consequences of technological mishaps such as TMI and Love Canal. Details of the TMI accident and our study are reported elsewhere (see, for example, Baum, 1990). Briefly, on 28 March 1979 the failure of a cooling system at the nuclear power plant resulted in the exposure of one of the radioactive cores. This led to the release of radioactive gas and water and eventually to an advisory that pregnant women and young children should evacuate the area. The accident received a great deal of media attention and conflicting information from utility and government officials led to confusion, mistrust and fear among area residents. The amount of radiation released during and after the accident as well as estimates of radiation to which area residents were exposed are still unclear.

Assessment of stress responding among area residents began with the sampling of residential areas within 5 miles of the damaged reactor. Quasi-random sampling in demographically comparable neighborhoods produced comparable groups of TMI area and control residents. Data collection always took place in the subjects' homes. Three different control groups were recruited initially, one of people living within 5 miles of an undamaged nuclear power plant, another, of people living within 5 miles of a coal-fired power plant and a third, of people living more than 5 miles from any power plant. Sampling and data collection for the control samples was similar to that used for the TMI residents.

Stress is difficult to define and measure and we have, as a result, developed an assessment strategy in which multiple markers of stress responding, including self-report, behavioral, psychophysiological and neuroendocrine response, are examined concurrently (Baum *et al.*, 1982). Self-report measures include assessments of symptom reporting, depression, anxiety, perceived distress and perceived threat. Behavioral measures include observational measures and administration of stress-sensitive performance tasks which require a great deal of concentration, persistence and/or frustration tolerance. Psychophysiological measurements in the TMI studies included examination of physicians' records along with repeated assessments of blood pressure and heart rate. Finally, 15 hour urine samples were collected repeatedly over time to examine neuroendocrine indexes of stress responding, namely urinary catecholamines (epinephrine and norepinephrine) and cortisol. Recognition that psychological stress may influence immune system functioning also led to the eventual examination of immune status in this population, which included the quantitative assessment of lymphocyte and leukocyte subpopulations and antibody titers to latent viruses (McKinnon *et al.*, 1989).

Symptoms of chronic stress, including elevations in symptom reporting, compromised task performance, elevations in blood pressure and urinary catecholamines and changes in immune system functioning were observed in the TMI area group for up to six years following the accident (Davidson *et al.*, 1982; Baum *et al.*, 1983a; Gatchel *et al.*, 1985; Davidson and Baum, 1986; McKinnon *et al.*, 1989; Baum, 1990).

### *14.3.1 CONTROL*

Early on, it was recognized that the victimization associated with technological disasters or hazards differed from that associated with natural disasters on several important dimensions (Baum *et al.*, 1983a). One such dimension is perceived control, which essentially reflects the belief that one can influence events or outcomes. In the case of natural disasters, expected and perceived control are typically low because natural forces are assumed to be unpredictable and uncontrollable. Tornadoes, earthquakes and floods often cause devastation and loss and advances in meteorology and seismology have enhanced our ability to predict these events. We can affect the likelihood of harm by evacuating or taking shelter, but it is unlikely that humans will ever be able to control disasters' occurrence. In contrast, expected and perceived control over the functioning of technological systems should be high, given the fact that humans created and developed these energy sources or products in an attempt to enhance quality of life. As Baum and Fleming noted, 'one of the ironies of twentieth century life is that the vast technological systems that we have created to tame the hostile world of our ancestors have turned and bitten the hand that feeds them' (1993: 665). As a result, technological disasters appear to directly threaten our perceived control because they represent a **loss of control** over a system that we created. This loss of control over a technology which 'should be' controllable may threaten perceptions of control in other areas of the victims' lives and lead to overall feelings of helplessness.

Findings from our research at TMI have suggested that the loss of control associated with the accident and its aftermath was associated with generalized feelings of helplessness, as TMI residents reported more helplessness and less perceived ability to control events in their lives than did comparison subjects (Davidson *et al.*, 1982). This perceived loss of control appeared to partially mediate the effects of the accident on long-term stress responding: TMI residents reporting lower levels of perceived control also exhibited greater symptom distress, performed more poorly on performance tasks and had higher levels of urinary stress-related hormones than did TMI residents who maintained higher levels of perceived control in the face of the TMI accident and aftermath (Davidson *et al.*, 1982).

A second characteristic of human-made disasters, frequently linked to the variable of perceived control, is the ready availability of a culpable agent upon which blame may be placed. Ordinarily, natural disasters are not attributable to such agents, again because people do not expect anyone to have control over or assume responsibility for destructive natural phenomena. Consequently, there is usually no individual, company, or corporation to which blame may be assigned. In contrast, radiation leaks, chemical spills and toxic landfills are defects or breakdowns in our machines and technology and as human-made hazards, may provide a focus for blame beyond the individual victim. It has been suggested that assigning blame for victimization to external sources may compromise the degree to which an individual perceives that they can control the negative events that befall them (Bulman and Wortman, 1977).

### *14.3.2 BLAME*

Given that TMI area residents reported less perceived control over events in their lives than did comparison subjects, we decided to look at the construct of blame (assumption of responsibility or assigning it to others) and to examine whether this variable was also related to stress responding (Baum *et al.*, 1983a). In such a case, self-blame should enhance feelings of perceived control by maintaining the belief that one caused one's own misfortune (or at least had a hand in it) and that this heightened feeling of control should be linked to less stress responding. On the other hand, victims of toxic exposure who are less likely to assume blame for the exposure itself and the degree to which subjects assigned responsibility for problems caused by the disaster to other people should be associated with diminished perceptions of control and greater distress. Results indicated that TMI subjects who accepted some responsibility for the problems caused by the disaster exhibited fewer symptoms of chronic stress than those who did not assume any responsibility. Specifically, TMI area residents who 'blamed' themselves for some of their disaster-related problems reported fewer somatic and psychological symptoms, performed better on a task requiring concentration and had lower levels of urinary epinephrine than did subjects who attributed responsibility to others. Those TMI subjects who accepted some blame did not differ from control subjects on stress-related outcome measures, suggesting a strong effect of perceived control and assignment of responsibility.

### *14.3.3 TOXIC EXPOSURE*

Another contrast between natural and human-made hazards is based on the fact that human-made hazards often involve potential exposure to toxic chemicals or radiation. This is, of course, most critical for assessing the impact of poisoned ecosystems on quality of life or health. Perceived exposure to toxins can be associated with long-term uncertainty about future health consequences for oneself or one's children. Health effects of exposure to radiation, pesticides, or chemicals leaking from a landfill may be slow to develop and the difficulty of objectively assessing actual exposure along with suspicion, fear and mistrust on the part of victims may all contribute to long-term stress. The heightened potential for cancer or genetic damage associated with toxic exposure is particularly important to consider, as these are among the most feared health outcomes in society (Wandersman and Hallman, 1993).

It has been reported that cancer is the most common fear following a technological mishap and it may be that the uncertainty and fear of cancer leads victims to interpret distressing psychological and/or physical symptoms as 'ominous warning signs of cancer' (Wandersman and Hallman, 1993: 684). While we have not directly assessed fear of cancer, we have examined the perception that the TMI accident posed a threat to one's health and the health of one's family (Gatchel *et al.*, 1985). Threat perception, both relative to self and to family, was consistently and significantly higher among TMI subjects than among controls (Gatchel *et al.*, 1985).

### *14.3.4 STRESS AT THREE MILE ISLAND*

These and subsequent studies have provided evidence of chronic stress responding in TMI residents. Though individual measures did not always produce significant differences, across more than 14 measurements over six years, TMI area subjects reported more somatic symptoms, greater depression and anxiety, more perceived threat, performed more poorly on behavioral tasks and had higher levels of blood pressure, urinary catecholamines and cortisol, following the accident, relative to control subjects (Gatchel *et al.*, 1985; Baum, 1990; Baum and Fleming, 1993). In addition to the mediating influences of perceived control and blame noted above, social support, coping style and intrusive imagery (i.e. frequent or intense recurrent thoughts about the accident) were also related to chronic stress in the TMI sample (Fleming *et al.*, 1982 Baum *et al.* 1983a; Baum, 1990). Those with more social support, those who used more emotion-oriented coping strategies and those who experienced fewer intrusive thoughts and images exhibited lower levels of chronic stress, particularly at assessments made five to six years after the accident (Baum, 1990). The emerging field of psychoneuroimmunology has provided growing evidence that stress may influence the functioning of the immune system and affect susceptibility to illness (e.g. Ader, 1981; Kiecolt-Glaser *et al.*, 1986, 1987; Cohen *et al.*, 1991; Herbert and Cohen, 1993). Consistent with this are findings from a study of a sub-sample of TMI and control subjects examining chronic stress and immune system functioning (McKinnon *et al.*, 1989). Conducted more than six years after the accident, blood samples were subjected to quantitative assessments of lymphocyte and leukocyte sub-populations and antibody titers to latent viruses. Urinary catecholamine and cortisol levels were also assessed. Consistent with predictions and previous findings, TMI subjects exhibited higher levels of urinary epinephrine than did control subjects and this hormonal variable was related to several immune indicators. The group of TMI residents showed significantly higher levels of neutrophils, lower levels of B-lymphocytes, T-suppressor/cytotoxic lymphocytes and natural killer cells and had higher antibody titers to two latent viruses, relative to control subjects. These findings suggest some long-term effects on immune system functioning in TMI subjects, particularly of adaptive, cell-mediated immunity. However, no evidence of clinical consequences of the immune system differences were found.

This brief review of some of our research at TMI illustrates the complexity of assessing chronic stress as a consequence of perceived exposure to an environmental threat and the powerful evidence of long-term stress responding among some of the TMI residents. In an attempt to replicate these findings in other populations, we have conducted longitudinal studies in communities exposed to a number of technological mishaps such as toxic chemical spills, contamination of a pesticide storage facility, hazardous leaking landfills and industrial exposure to toxic chemicals (Davidson *et al.*, 1986; O'Keeffe *et al.*, 1989; Baum *et al.*, 1992; Baum and Fleming, 1993). While assessment of immunocompetence remains to be investigated in these populations, our studies have consistently shown evidence of chronic stress among those experiencing toxic accidents or disasters relative to control samples. For example, elevations in blood pressure and urinary catecholamines, compromised task performance and symptom distress have been seen among subjects living near toxic hazards compared to 'less affected' control samples living more than 5 miles from any identified hazard (O'Keeffe *et al.*, 1989). This work is in progress, but data from one site, a landfill identified by the Environmental Protection Agency as one of the ten most hazardous landfills in the country, are remarkably similar to our findings from TMI. Long-term effects included elevations of somatic distress, psychological symptoms such as anxiety and depression and the psychophysiological and behavioral indexes of chronic stress mentioned previously (Davidson *et al.*, 1986). This research has also confirmed the importance of mediating variables such as loss of control and helplessness in determining long-term stress responding.

### *14.3.5 PERCEIVED THREAT AND STRESS*

In an attempt to specifically examine the importance of perceived threat of toxic exposure in long-term stress responding, independent of characteristics of human-made hazards such as loss of control, helplessness and blame, we have also examined residents of communities affected by radon gas problems. Prior studies of populations exposed to various technological hazards demonstrated that perceived toxic exposure is associated with concerns about long-term health implications. The threat of future health consequences associated with exposure to a potentially carcinogenic, mutagenic and/or teratogenic agent illustrates another attribute of technological mishaps that sets them apart from most natural hazards – namely chronicity. With most natural hazards, victims typically know when threat to life and property has passed. While it may take years to recover from the losses associated with the event, there is usually a clear point at which one can say that the worst is over and things will get better bit by bit. The threat posed by technological mishaps differs from this not only in that the feared health consequences may include cancer or genetic damage, but also in that the perceived threat persists long beyond the initial exposure.

To extend findings suggesting that human-caused hazards are associated with more evidence of chronic stress than are natural disasters, we conducted a two-year longitudinal study that assessed residents of communities exposed to a naturally occurring hazard, radon gas. Radon is a colorless, odorless radioactive gas that may pose a serious risk of lung cancer for those who reside in high-radon areas of the country (BEIR IV, 1988). Though naturally occurring, radon has some qualities we typically associate with technological disasters, including uncertainty about long-term health consequences, association with potential cancer risk and, as is the case with many human-made hazards, chronicity due to the lack of a clear point at which victims can acknowledge that the worst is over.

Preliminary results confirmed that subjects living in the high radon areas perceived it as a significant threat to their health. Those respondents living in areas where moderate to high radon was commonly found reported more perceived health threat than did control subjects. However, findings revealed that these subjects did not exhibit evidence of chronic stress and were comparable on most stress measures to controls. Task performance, blood pressure and urinary catecholamine levels in the radon subjects were comparable to those of control subjects over the course of the two-year study.

These results suggest that perceived threat may not be the most influential factor in determining chronic stress following exposure to an environmental hazard. Other characteristics of technological hazards, such as the presence of a culpable agent upon which to focus anger and blame and/or the perceived loss of control over a technology that should be controllable may be responsible in part for the chronic stress observed in the TMI and the other human-made hazard samples we have studied. However, it is unlikely that the differential response to toxic exposure in radon versus technological hazard samples can be explained by one or two variables. The relationship among variables such as perceived threat, blame, uncertainly, helplessness and intrusive imagery is complex and future research will need to address how these and other important constructs interact to determine who experiences chronic stress as a result of perceived toxic exposure.

## 14.4 CONSEQUENCES OF EXPOSURE TO ENVIRONMENTAL TOXINS

The research described above provides evidence that exposure to environmental hazards may cause chronic stress. The extent to which chronic stress has direct implications for long-term health outcomes is less clear. Our indicators of chronic stress include elevations in psychological symptoms, somatic distress, blood pressure, urinary catecholamines and cortisol, as well as poorer performance on tasks requiring concentration and persistence. In the TMI sample these chronic stress indicators, along with changes in immune system functioning, were observed long after the initial accident.

The clinical significance of this chronic stress, in terms of both mental and physical health, may not be fully understood for several years. In terms of psychological well-being, TMI area residents demonstrated elevations in anxiety, depression and symptoms of Post Traumatic Stress Disorder (PTSD), but prolonged elevations in severe psychopathology have not been identified (Davidson and Baum, 1986; Bromet *et al.*, 1990). Examination of 'high risk' populations such as those with frequent and/or severe recurrent thoughts about the accident and mothers who were pregnant and/or had young children at the time of the accident suggests that these sub-groups may be particularly vulnerable to chronic distress (Davidson and Baum, 1986; Baum, 1990; Bromet *et al.*, 1990). Six years after the accident, TMI residents reporting relatively few intrusive thoughts were comparable to controls on outcome measures, while those who experienced higher levels of intrusive imagery continued to show evidence of chronic stress (Baum, 1990). Mothers who continued to perceive that TMI was dangerous were also more likely to experience anxiety and depression than non-threatened mothers more than three years after the accident (Bromet *et al.*, 1990).

Many of the symptoms of chronic stress that we have measured may have direct effects on physical health as well. Elevations of blood pressure, urinary catecholamines and cortisol that were found did not exceed normal levels, but these elevations were prolonged and may have long-term health effects that remain to be determined. It has been suggested, for example, that stress-related changes in blood pressure and neuroendocrine hormones play a role in the development of coronary heart disease (Krantz *et al.*, 1981; Schneiderman, 1987; Kaplan *et al.*, 1993). Long-term elevations in blood pressure and urinary catecholamines suggest prolonged activation of the sympathetic nervous system, which may be linked to arterial damage and atherogenesis (Manuck *et al.*, 1989; Kaplan *et al.*, 1993). Likewise, we have found evidence suggestive of a less active immune system six years after the accident at TMI (McKinnon *et al.*, 1989). This is consistent with an accumulating body of evidence that suggests that psychological distress is reliably associated with compromised immune system functioning (see O'Leary, 1990; Weisse, 1992; and Herbert and Cohen, 1993, for reviews). The clinical significance of our immune findings are unclear, but values for leukocyte subpopulations remained within normal ranges and evidence of increased susceptibility to illness is not available. Increases in cancer rates among TMI residents have been observed and are reportedly linked to stress and perceptions of risk rather than increased exposure to radiation (Hatch *et al.*, 1991). Thus the chronic stress associated with exposure to a human-made environmental hazard may have direct implications for health. However, the clinical significance of these potential health threats remains to be determined definitively.

It is also likely that the chronic stress and exposure to toxins interact in complex ways to determine the ultimate effects of exposure to environmental hazards. Anderson *et al.*(1994) have proposed a model by which the chronic stress associated with a cancer diagnosis may interact with malignancy to influence cancer progression. These investigators point to stress-related changes in health behaviors such as diet, alcohol, smoking, sleep, exercise and adherence to medical recommendations that may influence cancer progression in a number of ways. They also suggest that stress-related changes in immune system functioning may influence cancer progression by influencing immunosurveillance and metastases. While the Anderson *et al.* (1994) model remains to be tested, it illustrates the indirect and direct ways by which the chronic stress associated with exposure to environmental hazards may interact with a carcinogenic, mutagenic and/or teratogenic substance to influence health outcomes.

Particularly relevant to such an interaction are data from a number of animal studies conducted by Seligman and his colleagues examining the role of perceived control in tumor development (Visintainer *et al.*, 1982; Seligman and Visintainer, 1985). In an initial study, rats were injected with live tumor cells and then randomly exposed either to escapable shock, inescapable shock, or no shock (Visintainer *et al.*, 1982). Animals exposed to inescapable shock were yoked to the animals exposed to escapable shock, so that the pattern, intensity and duration of shock exposure was identical in these two groups. The only difference was that one group (escapable shock) could terminate the shock by pressing a bar while the other group could not. They found that rats who had no control over the shock stressor (inescapable shock) were much more likely to develop tumors than rats in the escapable shock or control conditions, which were comparable. Specifically, 73% of the rats in the no-escape group developed tumors, compared to 48% of the escapable shock group and 49% of the no-shock control group. In explaining the mechanism by which control influenced tumor rejection, these investigators refer to several studies that demonstrate greater compromise of immunity following exposure to uncontrollable compared to controllable stressors (Maier *et al.*, 1985; Seligman and Visintainer, 1985).

In an extension of this study, Seligman and Visintainer (1985) examined the effects of an early experience with 'helplessness' (i.e. exposure to an uncontrollable stressor) on later tumor development. Weanlings (i.e. rats 27 days of age) were randomly exposed to the three conditions described above – escapable shock, non-escapable shock, or no shock. As young adults (63 days later), these same rats were injected with live tumor cells and then again exposed to one of the three experimental conditions, resulting in nine experimental conditions. Results indicated that the weanling shock condition did not influence tumor development in animals that were not exposed to shock later. However, among animals who were later exposed to either escapable or non-escapable shock, early shock experience had a significant influence on tumor development. Those with early experience with mastery (i.e. early exposure to escapable shock) were much less likely to develop tumors than were animals given an early experience with helplessness (i.e. early exposure to inescapable shock), regardless of whether later shock exposure was escapable or not. The authors argued that the early experience of having control over the shock stressor protected animals against subsequent loss of control and stress-related vulnerability to the injected tumor cells.

In our studies of TMI residents and residents of communities located near toxic landfills, helplessness and perceived loss of control have emerged as important mediators of stress responding (Baum and Fleming, 1993; Davidson *et al.*, 1982, 1986). Subjects in these studies consistently reported feeling that they had less control over events than did non-TMI samples and those reporting the highest levels of helplessness (i.e. lowest perceived control over events) also exhibited the greatest amount of chronic stress. If the work by Seligman and colleagues is applicable to humans, it suggests that the loss of control associated with this type of victimization may have important health consequences. It may be that the mechanisms by which helplessness increases susceptibility to injected tumor cells may also lead to an increased susceptibility to radiation or toxic chemical exposure.

## 14.5 CONCLUSIONS

The research and policy implications of this research on chronic stress following technological accidents have been outlined by Baum and Fleming (1993). Among these are the conviction that risk assessments must include consideration of the effects of stress in addition to the effects of the chemicals or radioactive substances potentially involved. Future research should also look more carefully at how stress and exposure may interact to influence long-term health outcomes. Although many of the dimensions on which toxic hazards are evaluated include impact on health, well-being and psychological contributions to health are rarely considered. We suggest that perception of a poisoned world, when associated with belief that one has been or may be exposed to hazards, can evoke stress and stress-related consequences for health and happiness.

Of particular interest to us is the possibility that stress and perceptions of toxic exposure may actually affect the impact of actual exposures. Estimates of health effects of toxic leaks or accidents are usually based on studies of different levels of exposure in normally functioning organisms. They are rarely if ever based on studies of these exposures on stressed organisms, **even though stress alters every organ system in the body and changes the acute or chronic milieu in which toxic exposure occurs.**

Stress, among other things, appears to stimulate liver activity and may actually increase hepatic clearance of toxins that are degraded there. It has been suggested that stress affects urinary tract characteristics and increases voiding of nicotine, resulting in heightened smoking (Schachter, 1978). Similarly, stress appears to slow absorption and increase detoxification of alcohol in experimental situations, suggesting a mechanism by which stress may increase alcohol consumption (Breslin *et al.*, 1994). More accurate impressions of the impact of drugs of abuse, toxic exposure, or other environmental insults may depend on more systematic consideration of organismic states and their interaction with chemical or radiation exposure.

While we have considered a range of issues related to the consequences of perceptions of toxic contamination and potential exposure, we recognize the difficulties inherent in studying and ameliorating contaminating conditions in this manner. However, the burdens associated with stress and other psychologically mediated aspects of toxic hazards, together with perceptions of risk, constitute a powerful mediator of health and well-being as well as an important aspect of public health and environmental health sciences.

## REFERENCES


Abelson, P. H. (1992) Remediation of hazardous waste sites. *Science*, **255**, 901.

Ader, R. (ed.) (1981) *Psychoneuroimmunology*, Academic Press, New York.

Anderson, B. L., Kiecolt-Glaser, J. K. and Glaser, R. (1994) A biobehavioral model of cancer stress and disease course. *American Psychologist*, **49**, 389–404.

Baum, A. (1990) Stress, intrusive imagery, and chronic distress. *Health Psychology*, **9**, 653–75.

Baum, A. and Fleming, I. (1993) Implications of psychological research on stress and technological accidents. *American Psychologist*, **48**, 665–72.

Baum, A. Fleming, R. and Singer, J. E. (1983a) Coping with victimization by technological disaster. *Journal of Social Issues*, **39**, 117–38.

Baum. A., Gatchel, R. J. and Schaeffer, M. A. (1983b) Emotional, behavioral, and psychological effects of chronic stress at Three Mile Island. *Journal of Consulting and Clinical Psychology*, **51**, 565–72.

Baum, A., Grunberg, N. E. and Singer, J. E. (1982) The use of psychological and neuroendocrinological measurements in the study of stress. *Health Psychology*, **1**, 217–36.

Baum, A., Fleming, I., Israel, A. and O'Keeffe, M. K. (1992) Symptoms of chronic stress following a natural disaster and discovery of a human-made hazard. *Environment and Behavior*, **24**, 347–65.

Becker, M. H. (1993) A medical sociologist looks at health promotion. *Journal of Health and Social Behavior*, **34**, 1–6.

BEIR IV (Committee on the Biological Effects of Ionizing Radiations) (1988) *Health Risks of Radon and Other Internally-Deposited Alpha-Emitters*, National Academy Press, Washington, DC.

Breslin, F. C., Hayward, M. and Baum, A. (1994) Effect of stress on perceived intoxication and the blood alcohol curve in men and women. *Health Psychology*, **13**, 479–87.

Bromet, E. J., Parkinson, D. K. and Dunn, L. O. (1990) Long term health consequences of the accident at Three Mile Island. *International Journal of Mental Health*, **19**, 48–60.

Bulman, R. J. and Wortman, C. B. (1977) Attribution of blame and coping in the 'real world': severe accident victims react to their lot. *Journal of Personality and Social Psychology*, **35**, 351–63.

Cohen, S., Tyrell, D. A. and Smith, A. P. (1991) Psychological stress and susceptibility to the common cold. *New England Journal of Medicine*, **325**, 606–12.

Davidson, L. M. and Baum, A. (1986) Chronic stress and posttraumatic stress disorders. *Journal of Consulting and Clinical Psychology*, **54**, 303–8.

Davidson, L. M., Baum, A. and Collins, D. L. (1982) Stress and control related problems at Three Mile Island. *Journal of Applied Social Psychology*, **12**, 349–59.

Davidson, L. M., Fleming, I. and Baum, A. (1986) Post-traumatic stress as a function of chronic stress and toxic exposure, in *Trauma and Its Wake* (ed. C. P. Figley), Brunner/Mazel, New York, pp. 55–77.

Fleming, R., Baum, A., Gisreal, M. M. *et al.* (1982) Mediating influence of social support on stress at Three Mile Island. *Journal of Human Stress*, **8**, 14–22.

Gatchel, R. J., Schaeffer, M. A. and Baum, A. (1985) A psychophysiological field study of stress at Three Mile Island. *Psychophysiology*, **22**, 175–81.

Hatch, M. C., Wallenstein, S., Beyea, J. *et al.* (1991) Cancer rates after the Three Mile Island nuclear accident and proximity of residence to the plant. *American Journal of Public Health*, **81**, 719–24.

Herbert, T. B. and Cohen, S. (1993) Stress and immunity in humans: a meta-analytic review. *Psychosomatic Medicine*, **55**, 364–79.

Houts, P. S., Tokuhata, G. K., Bratz, J. *et al.* (1991) Effect of pregnancy during TMI crisis on mothers' mental health and their child's development. *American Journal of Public Health*, **81**, 384–6.

Kaplan, J., Manuck, S. B., Williams, J. K. and Stawn, W. (1993) Psychosocial influences on atherosclerosis: evidence for effects and mechanisms in nonhuman primates, in *Cardiovascular Reactivity to Stress and Disease* (eds J. Blascovich and E. S. Katlin), American Psychological Association, Washington, DC, pp. 3–26.

Kiecolt-Glaser, J. K., Glaser, R., Dyer, C., Shuttleworth, E. C., Ogrocki, P., and Speicher, C. E. (1987) Chronic stress and immunity in family care-givers of Alzheimers disease victims. *Psychosomatic Medicine*, **49**, 523–35.

Kiecolt-Glaser, J. K., Glaser, R., Strain, E. *et al.* (1986) Modulation of cellular immunity in medical students. *Journal of Behavioral Medicine*, **9**, 5–21.

Kohut, A. and Schriver, J. (1989) *The Environment* (Report No. 285). The Gallup Report, Princeton, NJ.

Korgeski, G. P. and Leon, G. R. (1983) Correlates of self-reported and objectively determined exposure to Agent Orange. *American Journal of Psychiatry*, **140**, 1443–9.

Krantz, D. S., Schaeffer, M. A., Davia, J. E. *et al.* (1981) Extent of coronary atherosclerosis, type A behavior, and cardiovascular response to social interaction. *Psychophysiology*, **18**, 654–64.

Maier, S. F., Laudenslager, M. L. and Ryan, S. M. (1985) Stressor controllability, immune function, and endogenous opiates, in *Affect, Conditioning, and Cognition: Essays on the Determinants of Behavior* (eds F. R. Brush and J. B. Overmier), Erlbaum, Hillsdale, NJ, pp. 183–201.

Manuck, S. B., Muldoon, M. F., Kaplan, J. R. *et al.* (1989) Coronary artery atherosclerosis and cardiac response to stress in cynomolgus monkeys, in *In Search of Coronary Prone Behavior: Beyond Type A* (eds A. W. Siegman and T. M. Dembroski), Erlbaum, Hillsdale, NJ, pp. 207–27.

McKinnon, W., Weisse, C. S., Reynolds, C. P. *et al.* (1989) Chronic stress, leukocyte subpopulations, and humoral response to latent viruses. *Health Psychology*, **8**, 389–402.

McPherson, M. (1984) *Long Time Passing*, Doubleday, Garden City, NY.

O'Keeffe, M. K. and Baum, A. (1990) Conceptual and methodological issues in the study of chronic stress. *Stress Medicine*, **6**, 105–15.

O'Keeffe, M. K., Baum, A. Weiss, L. and Davidson, L. (1989) Chronic Stress in Victimized Communities. Paper presented at the convention of the American Psychological Association, New Orleans, LA.

O'Leary, A. (1990) Stress, emotion, and human immune function. *Psychological Bulletin*, **108**, 363–82.

Poland, A. and Glover, E. (1973) 2,3,7,8-Tetrochlorodibenzo-p-dioxin: a potent inducer of aminolevulinic acid synthetase. *Science*, **179**, 476–7.

Poland, A. and Kimbrough, R. D. (eds) (1984) *Biological Mechanisms of Dioxin Action*, Banbury Report, **18**, Cold Spring Harbor Laboratory, Cold Spring Harbor, NY.

Quarantelli, E. L. (1985) What is disaster? The need for clarification in definition and conceptualization in research, in *Disasters and Mental Health: Selected Contemporary Perspectives* (ed. B. J. Sowder) (DHHS Publication No. ADM 85–1421), National Institute of Mental Health, Rockville, MD, pp. 41–73.

Schachter, S. (1978) Pharmacological and psychological determinants of smoking. *Annals of Internal Medicine*, **88**(1), 104–14.

Schneiderman, N. (1987) Psychophysiologic factors in atherogenesis and coronary artery disease. *Circulation*, **76**, (Suppl. I), 41–7.

Seligman, M. E. P. and Visintainer, M. A. (1985) Tumor rejection and early experience of uncontrollable shock in the rat, in *Affect, Conditioning, and Cognition: Essays on the Determinants of Behavior* (eds F. R. Brush and J. B. Overmier), Erlbaum, Hillsdale, NJ, pp. 203–10.

Sowder, B. J. (ed.) (1985) *Disasters and Mental Health: Selected Contemporary Perspectives* (DHHS Publication No. ADM 85–1421), National Institute of Mental Health, Rockville, MD.

Stone, R. (1992) I.O.M. to study Agent Orange. *Science*, **257**, 1335.

Tyson, R. (1992) Poll: environment tops agenda. *USA Today*, 1 June, p. 1.

US Committee on Government Operations (1993) *The Health Risks of Dioxin: Hearing Before the Human Resources and Intergovernmental Relations Subcommittee*, US Government Printing Office, Washington, DC.

US Committee on Veterans' Affairs (1994) *National Academy of Sciences Report on Health Effects of Agent Orange*, US Government Printing Office, Washington, DC.

Visintainer, M. A., Volpicelli, J. R. and Seligman, M. E. P. (1982) Tumor rejection in rats after inescapable or escapable shock. *Science*, **216**, 437–9.

Wandersman, A. H. and Hallman, W. K. (1993) Are people acting irrationally? Understanding public concerns about environmental threats. *American Psychologist*, **48**, 681–6.

Weisse, C. S. (1992) Depression and immunocompetence: a review of the literature. *Psychological Bulletin*, **111** (3), 475–88.

Wise, M. Z. (1991) UN report blames stress, not radiation, for Chernobyl illnesses. *The Washington Post*, 22 May, p. A24.

# PART SIX

## *Permeation into Literature*

15

# *Ecocriticism: literary studies in an age of environmental crisis*

CHERYLL GLOTFELTY

This book aims to make connections between two groups that have worked in relative isolation – human health scientists on the one hand, and ecosystem health scientists on the other. While bridging the gap between human and ecosystem sciences is clearly a step in the right direction, there exists a vastly wider chasm which divides all the sciences from the humanities. As C. P. Snow argued in his renowned lecture entitled 'The Two Cultures and the Scientific Revolution' (1959), 'Literary intellectuals at one pole – at the other scientists . . . Between the two a gulf of mutual incomprehension.'

One measure of this gulf is the great distance between the professional languages of scientists and literary intellectuals. Most literary scholars, for example, would be unable to decipher the following sentence, one which presumably is readily comprehensible to both human and ecosystem scientists: 'To address this question, we first used transfection assays to examine DNA isolated from the molluscan tumors for the presence of activated oncogenes' (Van Beneden, 1993). Similarly, I suspect that the following sample sentence from a respected literary critic would baffle most scientists: 'The story's self-exemptions from internalized norms of major pastoral performance clearly require an overdetermined differential praxis' (Renza, 1984: 133). Developing the ability to make meaning out of sentences like these can require years of specialized training.

It is not realistic to expect that scientists and literary scholars master one another's professional jargon; none the less, it might be both possible and productive for the two different cultures to engage in a conversation about a common concern, in this case, human and ecosystem health. Accordingly, in this chapter, I shall review for the benefit of scientists some of the ways in which the profession of English is developing 'ecological' approaches to the study of

*Interconnections Between Human and Ecosystem Health.* Edited by Richard T. Di Giulio and Emily Monosson. Published in 1996 by Chapman & Hall, London. ISBN 0 412 62400 1.

language and literature. Such approaches have recently coalesced under the umbrella term 'ecocriticism'. Simply defined, ecocriticism studies the interconnections between the physical world and human culture, specifically the cultural artifacts language and literature. Ecocriticism has one foot in literature and the other on land; it is a critical and theoretical discourse that negotiates between the human and the non-human.

It should be noted that literary scholars normally engage in both teaching and research. Although it is probably as teachers that English professors exert the most direct influence in the world, their teaching is informed by research. Hence, this review focuses on research, namely, literary criticism and theory. Discussions of teaching can be found in Waage (1985) and Glotfelty (1993a, 1993b), while Orr (1992) offers a provocative discussion of higher education in general.

Many literary scholars concerned about the twentieth century environmental crisis initially find themselves in a quandary: while their temperaments and talents have deposited most of them in English departments, as environmental problems mount, they wonder how they can contribute to planetary health from within their capacity as professors of literature. Answers are at first elusive. After all, what does correct comma placement have to do with chlorofluorocarbons? How can analyzing Shakespeare shed light on deforestation? What is the relationship between river symbolism in *Huckleberry Finn* and toxic waste in the Mississippi Delta?

At first it may appear that the world of literary studies and the world of environmental realities exist in two different universes. Yet, if we agree with Barry Commoner's first law of ecology, that 'Everything is connected to everything else', we must ultimately conclude that literature does not float above the physical world in some rarified, aesthetic ether, but, rather, it plays a part in an immensely complex global system, in which energy, matter *and ideas* interact in a perpetual ebb and flow. Put simply, literature acts on people, and people act on the world. Ecocritics generally agree that people have been and are presently acting unwisely in the world. (Litanies of environmental ills, from toxic and nuclear waste, to destruction of the rain forests, to loss of habitat are a hallmark of ecocriticism.) What ecocritics wonder is how literature may have influenced these harmful actions, how literature may bring about healthful actions, and what role literary criticism may play in the process.

Ecocritics and theorists ask questions like the following: How is nature represented in this sonnet? What role does the physical setting play in the plot of this novel? Are the values expressed in this play consistent with ecological wisdom? In what ways and to what effect is the environmental crisis evident in contemporary literature and popular culture? How do our metaphors of the land influence the way we treat it? In what ways has literacy itself affected humankind's relationship to the natural world? Do men write about nature differently than women do? What view of nature informs US Forest Service reports, and what rhetoric enforces this view? How has the concept of wilderness changed over time? What bearing might the science of ecology have on literary studies? How is science itself open to literary analysis? What cross-fertilization is possible between literary studies and environmental discourse in related disciplines, like history, philosophy, psychology, art history and ethics?

Before reviewing varieties of ecocriticism and theory, it might be helpful for scientists if I answer some basic questions. What is literary criticism? What is literary theory? And what might constitute ecological versions of these endeavors?

Simply put, literary criticism is scholarly writing about *literature*, while literary theory is scholarly writing about *literariness*. Criticism is concrete and discusses particular *texts*, while theory is abstract and discusses *textuality*. Criticism aims to illuminate and interpret specific works, while theory *uses* specific works to illustrate a larger argument about literature in general.

Since some theory of literature, whether consciously articulated or not, underlies all criticism, let us first take a closer look at theory. Literary theory, which asks very basic questions about the nature of writing and reading, about language and signification, may seem to be too abstract for an ecological approach. To demand that literary theory engage ecological issues may at first seem as preposterous as calling for the greening of pure mathematics. But let us reconsider.

Theory examines the relations between writers, literature, readers and the world. In most literary theory 'the world' is synonymous with society. Marxist literary theory, for example, studies how the class structure of society impinges upon the production and consumption of writing, while feminist literary theory looks at how language perpetuates a patriarchal social order. An ecological form of literary theory expands the notion of 'the world' to include the entire biosphere, asking questions like 'How does literature function within the ecosystem?' or 'How does a given textual representation of nature affect the way we treat actual nature?'

In order to characterize how literature interacts with the biosphere, one must attempt to characterize the relationship between literature and human life. Theorists have identified two basic patterns: literature as mirror and literature as model. Those who regard literature as a mirror of life hold that art imitates life, whereas those who see literature as a model for life believe that life imitates art. Either conception of literature may be cogent, depending especially on the work of literature under scrutiny, for some authors incline toward presenting the world as they see it, while others are more visionary and present the world as it might be. The issue is further complicated because readers sometimes model their own lives after works whose intention was to be mirrors; conversely, works whose aim was to be models, at some level cannot avoid *reflecting* the prevailing world view.

Despite the fact that most pieces of literature function both as mirrors and as models, ecocritical treatments of literature tend to adopt one or the other conception. I shall take them up consecutively and review the kinds of ecocritical work being done in each camp. Critics who regard literature as a mirror of society, reflective of its values, generally scrutinize literature with a critical eye, searching for clues for where we have gone wrong.

One example of this kind of criticism is Annette Kolodny's book *The Lay of the Land* (1975), which surveys a wide range of male-authored American literature, and finds that it pervasively employs metaphors that depict the American landscape as feminine, as virgin land or nurturing mother. Kolodny hypothesizes that because we have gendered the land female, the same patterns of aggression, domination and violation that can be found in men's treatment of women also obtain in our culture's treatment of the land. Thus, the metaphors we use to describe nature – 'fertile plains,' 'the fresh green breast of the New World,' 'the rape of the land' – may reveal the subconscious motives for our treatment of it. Critics like Kolodny hope that their efforts to identify environmentally pathological patterns in literature will bring those patterns to public consciousness so that they can be diagnosed and replaced with healthy ones. Another example of criticism that regards literature as a mirror or barometer of our culture is Scott Sanders's essay 'Speaking a word for nature (1987)' which surveys recent critically acclaimed fiction and laments the absence of nature-awareness in it, worrying that the lack of ecological consciousness in our best-sellers reflects an endemic myopia in our culture.

Turning now to ecocriticism that treats literature as a model, we recall that critics favoring this approach believe that people pattern their lives after the patterns they find in books; in this view, literature does not so much *reflect* the way things are as it *affects* them. Ecocritics subject these models to a literary version of Environmental Impact Statements, assessing whether the literature being studied, if imitated in life, would lead to ecologically *de*structive or *con*structive behavior. One of the more famous examples of this kind of criticism is Lynn White Jr's controversial 1967 essay 'The historical roots of our ecologic crisis'. White takes as his text the Bible, undeniably a model for conduct, advancing what has since become a familiar argument that Judeo-Christianity is damnably anthropocentric in its belief that 'God planned all of [physical creation] explicitly for man's benefit and rule: no item in the physical creation had any purpose save to serve man's purposes. And, although man's body is made of clay, he is not simply part of nature: he is made in God's image' (1967: 1205).

White and others trace many of the world's environmental ills to this Judeo-Christian worldview as codified in its sacred text. Such a view is anthropocentric rather than biocentric; it regards human beings as separate from the rest of creation rather than as connected to it; it fosters an attitude of arrogance rather than one of humility before the natural world; and it leads to senseless exploitation and domination rather than wise acceptance of limitations and compliance with natural laws.

Ecocritics as a group unite in condemning the anthropocentric worldview, with its attendant attitudes and actions. According to White and others, unhealthy attitudes like anthropocentrism beget harmful actions, which in turn create unhealthy conditions for all life. As a group, then, ecocritics support concepts such as interconnectedness and interrelationship, and values such as community and cooperation; most ecocritics advocate rootedness and affection for a particular place, one's home bioregion. Ecocritics promote literature that embodies these values, while they criticize literature that portrays humankind as separate from nature.

Whereas some ecocritics focus on specific texts, such as the Bible, or specific time periods, such as the contemporary, others turn their attention to classic literary modes. Joseph Meeker in his groundbreaking monograph *The Comedy of Survival: Studies in Literary Ecology* (1972) regards humankind as a 'literary animal' and evaluates the literary modes of tragedy and comedy on the basis of whether or not they tend to promote behavior consistent with the goals of survival and ecological harmony. Meeker denounces Greek tragedy (*Antigone* and others) for propounding three ideas which are ecologically catastrophic in their consequences: 'the assumption that nature exists for the benefit of mankind, the belief that human morality transcends natural limitations, and humanism's insistence upon the supreme importance of the individual personality' (1972: 52). In contrast, he applauds comedy (Aristophanes's *Lysistrata* is one example) for its endorsement of community and its affirmation of life:

> Comedy is concerned with muddling through, not with progress or perfection . . . The comic point of view is that high moral ideals and glorified heroic poses are themselves largely based upon fantasy and are likely to lead to misery or death for those who hold them. In the world as revealed by comedy, the important thing is to live and to encourage life even though it is probably meaningless to do so. (1972: 41)

Meeker, a human ecologist with a PhD in comparative literature and postdoctoral studies in wildlife ecology and ethology, further speculates that

> If comedy is essentially biological, it is possible that biology is also comic. Some animal ethologists argue that humor is not only a deterrent to aggression, but also an essential ingredient in the formation of intraspecific bonds. It appears to have a phylogenetic basis in many animals as well as in man. Beyond this behavioral level, structures in nature also reveal organizational principles and processes which closely resemble the patterns found in comedy. Productive and stable ecosystems are those which minimize destructive aggression, encourage maximum diversity, and seek to

> establish equilibrium among their participants – which is essentially what happens in literary comedy. Biological evolution itself shows all the flexibility of comic drama and little of the monolithic passion peculiar to tragedy. (1972: 41)

While Meeker examines classic literary modes, other ecocritics turn their attention to the non-fiction genre of nature writing, as exemplified by writers like John James Audubon, Henry David Thoreau, John Muir, Mary Austin, Joseph Wood Krutch, Edwin Way Teale, Henry Beston, Rachel Carson, Edward Abbey, Annie Dillard, Ann Zwinger, Barry Lopez and Terry Tempest Williams, to name only the most famous. Some critics, like Glen Love in his influential essay 'Revaluing nature: toward an ecological criticism' (1990), believe that this 'nature-oriented literature' offers a needed antidote to our narrowly anthropocentric view of life; nature writing shows regard for the non-human and privileges 'eco-consciousness' over 'ego-consciousness' (1990: 205). Ecocritics hope that by drawing attention to nature writing, this relatively neglected genre will gain respect, be more widely read and taught, and will help bring about the change in consciousness necessary to improve human mental health and to heal our fractured relationship with the natural world.

In addition to examining literary works, periods, modes and genres, ecocritics search for authors whose work manifests ecological vision. William Rueckert, in an extremely interesting, idiosyncratic and brilliant essay entitled 'Literature and ecology: an experiment in ecocriticism' (1978), is the first person to coin the term 'ecocriticism' to describe his attempt to develop an 'ecological poetics', in which the science of ecology, with its concept of the ecosystem and its emphasis on interconnections and energy flow, serves as a metaphor for how poetry functions in society. He draws attention to writers as diverse as Gary Snyder, Adrienne Rich, W. S. Merwin, Walt Whitman, William Faulkner, Henry David Thoreau and Theodore Roethke for their ecological vision. Rueckert believes that the work of such writers could have a purgative-redemptive effect on humanity and that the critic's role is to act as mediator to help 'release the energy and power stored in poetry so that it may flow through the human community' (1978: 74) and be translated into social, economic, political and individual programs of action.

As I hope this review has demonstrated, literary scholars are making a substantial effort to contribute toward understanding humanity's place in the ecosystem. From the beginning of recorded history, literature has pondered humanity's place in the cosmos; today such questions tend to display an urgency which derives from knowledge of the environmental crisis, and they are additionally informed by concepts from the science of ecology. The efforts of ecocritics join those being made in other branches of the humanities as we all begin to recognize that current environmental problems are largely of our own making, are, in other words, a byproduct of culture. Historian Donald Worster explains:

> We are facing a global crisis today, not because of how ecosystems function but rather because of how our ethical systems function. Getting through the crisis requires understanding our impact on nature as precisely as possible, but even more, it requires understanding those ethical systems and using that understanding to reform them. Historians, along with literary scholars, anthropologists, and philosophers, cannot do the reforming, of course, but they can help with the understanding. (1993: 27)

Answering the call to understanding, Worster and other historians are writing environmental histories, studying the reciprocal relationship between humans and land, considering nature not just as the stage upon which the human story is acted out, but as an actor in the drama. They trace the complex dynamic among environmental conditions, economic modes of production, and cultural ideas through time. Anthropologists have long been interested in the connection between culture and geography; their work on primal cultures in particular may help the rest of us to respect those people's right to survive, and to learn about the value systems and rituals that help them live sustainably. In philosophy, various subfields like environmental ethics, deep ecology, ecofeminism and social ecology have emerged in an effort to understand and critique the root causes of environmental degradation and to formulate an alternative view of human nature that will provide an ethical and conceptual foundation for right relations with the earth.

Literary scholars specialize in questions of value, meaning, point of view and language, and it is in these areas that they are making a substantial contribution to environmental thinking. Believing that the environmental crisis has been exacerbated by our fragmented, compartmentalized and overly specialized way of knowing the world, humanities scholars are increasingly making an effort to educate themselves in the sciences and to talk across disciplinary fences. Scientists can teach literary intellectuals about a physical world much richer, more intricate and even more beautiful than anything conceived in the imagination. Literary intellectuals can teach scientists that the very language that they employ to describe the physical world is itself a product of culture and hence is inextricably tied to the world of ethics. Viewed holistically, health – both human and ecosystem health – is not just a state of matter but a condition of the spirit.

## REFERENCES

Glotfelty, C. (1993a) Teaching green: ideas, sample syllabi, and resources. *ISLE: Interdisciplinary Studies in Literature and Environment*, **1,** 151–78.

Glotfelty, C. (1993b) Western, yes, but is it literature?: teaching Ronald Lanner's *The Pinon Pine*'. *Western American Literature*, **27,** 303–10.

Kolodny, A. (1975) *The Lay of the Land: Metaphor as Experience and History in American Life and Letters*, University of North Carolina Press, Chapel Hill, NC.

Love, G. A. (1990) Revaluing nature: toward an ecological criticism. *Western American Literature*, **25,** 201–15.
Meeker, J. (1972) *The Comedy of Survival: Studies in Literary Ecology*, Scribner's, New York.
Orr, D. W. (1992) *Ecological Literacy: Education and the Transition to a Postmodern World*, State University of New York Press, Albany, NY.
Renza, L. A. (1984) *'A White Heron' and the Question of Minor Literature*, University of Wisconsin Press, Madison, WI.
Rueckert, W. (1978) Literature and ecology: an experiment in ecocriticism. *Iowa Review*, **9,** 62–86.
Sanders, S. R. (1987) Speaking a word for nature. *Michigan Quarterly Review*, **26,** 648–62.
Snow, C. P. (1959) The two cultures and the scientific revolution. *Encounter*, **12.6,** 17–24; **13.1,** 22–7.
Van Beneden, R. (1993) Abstract for the session on Interconnections Between Human and Ecosystem Health, SETAC Conference, Houston, TX.
Waage, F. O. (ed.) (1985) *Teaching Environmental Literature: Materials, Methods, Resources.* Modern Language Association, New York.
White, L., Jr (1967) The historical roots of our ecologic crisis. *Science*, **155,** 1203–7.
Worster, D. (1993) *The Wealth of Nature: Environmental History and the Ecological Imagination.* Oxford University Press, New York.

# 16 The literature of toxicity from Rachel Carson to Ana Castillo

TERRELL DIXON

## I

The recent growth in concern about environmental issues has made itself felt in many aspects of American academic life – an increase in environmental science courses and social science courses, an increase in humanities courses in such fields as history and philosophy, and rapid growth in the number of ecology and environmental studies programs and schools. Nowhere, however, has this developing cultural interest been more dramatically manifest than in the study of literature. From what was once a small group of scholars interested chiefly in the study of the history of American nature writing, the field, after a slow start in the early 1970s, has grown rapidly both in the breadth and depth of what is studied and in the number of its practitioners.

Calls for an ecologically oriented literary criticism can be said to begin with Joseph Meeker's *The Comedy of Survival: Studies in Literary Ecology* (1972), a topic and a book that merited, but failed to receive, much attention in the literary academy. More recently, however, critics such as Cheryll Burgess Glotfelty (1989), Karl Kroeber (1993) and Glen Love (1990) have again stated the need for an ecological literary criticism. In the climate of renewed environmental concern in the late 1980s and early 1990s, their statements have met with more success. Now usually designated as the study of literature and the environment, the field has developed a substantial scholarly association – the Association for the Study of Literature and the Environment (ASLE). This organization has attracted some 550 members in its first two years of existence and sponsored panels at almost every important conference on literary criticism; it will soon begin to hold biannual meetings of its own. Literary studies of the environment with an interdisciplinary reach have also

*Interconnections Between Human and Ecosystem Health.* Edited by Richard T. Di Giulio and Emily Monosson. Published in 1996 by Chapman & Hall, London. ISBN 0 412 62400 1.

been a prominent feature of conferences such as the annual North American Interdisciplinary Wilderness Conference, and increasingly, other professional societies, such as the Society for Environmental Toxicity and Chemistry where a short version of the essay reproduced in this chapter was first delivered, have sought to make connections with literary scholars across disciplinary lines.

Scholarly publication has grown as well. There now exist, for example, a substantial semi-annual newsletter, the *American Nature Writing Newsletter*, and a journal, *ISLE. Interdisciplinary Studies in Literature and the Environment*, which originated independently and have now been affiliated with ASLE. While the important books are too numerous to attempt even a short list here, it is clear that trade anthologies of American nature and environmental writing, classroom anthologies, scholarly essays and books also have become prominent features of the academic and literary landscape.

A focus on toxicity has not, however, been a significant part of this growing scholarly interest thus far. To fill that gap, this chapter examines how prose literature has dealt with the environmental problem of toxicity. In doing this, it traces, patterns, linkages and changes which characterize the response that literature has made to the growing problem of toxicity from Rachel Carson in 1962 to Ana Castillo (1993) (see Author's Note, p. 257). I shall try to develop this literary analysis so that it not only increases knowledge within the field of environmental literature, but so that it also helps foster understanding across other disciplines in the humanities and the sciences. While we, as scholars in the various fields of environmental studies, often stress ecological interconnectedness within our respective fields, we also, somewhat paradoxically, often keep our ideas about inclusiveness bound within the language and thus the readership of our individual fields of inquiry. We seldom advance, in any active way, a broader notion of wholeness extending across traditional academic lines.

This chapter attempts a modest, preliminary step in that direction. The initial analyses look directly at how literature and science come together in two important literary works, and the language used throughout this discussion of the literature of toxicity seeks to make its points in ways that will be accessible across disciplinary lines. The hope, then, is that this can be a start toward an interdisciplinary accessibility and connectedness that will help us work together to solve the problems of toxicity.

## II

In 1951, Rachel Carson published a three part series in *The New Yorker* called 'Profile of the Sea'. That series and her later books shaped the critical first stage in the contemporary literature of toxicity. Carson had begun her academic training years earlier as an English major at the Pennsylvania College for women (now Chatham College), but she changed her major to science, first to biology and then to zoology, the field in which she went on to do her MA at Johns Hopkins. At this early point in her life, she thought that science and literature were mutually exclusive endeavors, that she had to give up writing for science. 'I thought I had to be one or the other; it never occurred to me, or apparently anyone else that I could combine the two careers' (Brooks, 1989: 17).

When the *New Yorker* series was first published in book form as *The Sea Around Us* in 1951, it became clear that she could successfully combine the two. This book was, as Robert Gottlieb (1993: 81–6) has noted, an 'extraordinary success'. 'It stayed on the best seller list for eighty-six weeks, sold more than 2 million copies, and was translated into thirty-two languages.' It also won for its author a distinguished literary prize: the National Book Award. Her acceptance speech for this prize is as instructive today as when it was given.

> Many people have commented on the fact that a work of science should have a large popular sale. But this notion that 'science' is something that belongs in a separate compartment of its own, apart from everyday life, is one that I would like to challenge. We live in a scientific age; yet we assume that knowledge is the prerogative of only a small number of human beings, isolated and priest like in their laboratories. This is not true. The materials of science are the materials of life itself. Science is part of the reality of living; it is the what, the how, and the why in everything in our experience. It is impossible to understand man without understanding his environment and the forces that have molded him physically and mentally. (Brooks, 1989: 128)

What Rachel Carson talks about in this speech and what she exemplifies in her books is the ability to bridge the gap between what C. P. Snow (1963) called the 'two cultures', the sciences and the humanities. Since then, however, scientific research and humanities study have become even more specialized and more isolated from each other, making such ability both more rare and more necessary. As the distinguished scientist and writer E. O. Wilson has argued recently, the 'convergence of science and the humanities' is 'the grail of the academic community'. He also suggests, in the same talk recorded by Edward Lueders (1989: 9–13), that 'creativity in the two branches of learning might converge' most profitably in the area of nature writing. *The Sea Around Us* demonstrated that the nature writer Rachel Carson, working simultaneously as scientist and as writer, had both the creative capacity for that convergence and the ability to bring it to the attention of the larger world.

Her next and final book, *Silent Spring*, continues this effort, and it also initiates, for our time, what we can designate as the literature of toxicity, a literary theme that takes on increasing social and cultural importance in the second half of the twentieth century. *Silent Spring* was first published in 1962, at the time when questions about pollution and toxicity were just beginning to attract public attention. It is not too much to say, as Paul Brooks has suggested

(1989: xii), that her book was 'nothing less than an attempt to create a new environmental consciousness'. If this book is, by necessity of its subject, less lyrical and more critical than *The Sea Around Us*, it is still another remarkable blend of literary expression and scientific material.

Work toward this first contemporary book on toxicity began in 1958 with a good friend, Olga Owens Hutchins, describing the 'agonizing deaths' of birds in her area when state officials sprayed with DDT. At her friend's request that she help research this problem, Rachel Carson quickly became drawn into a full-scale exploration of chemical toxicity. Her research and her correspondence with such other researchers as Clarence Cottam, former assistant director of the Fish and Wildlife Service, M. M. Hargraves of the Mayo Clinic, and Wilhem C. Hueper of the National Cancer Institute, were, as Paul Brooks has noted, extensive.

Her collection of material occurred at a relatively innocent time in our culture's history, a time before disasters such as Love Canal and Three Mile Island saturated the public mind with the range and prevalence of toxic dangers. 'Better Things for Better Living Through Chemistry' was a very successful advertising phrase, pesticides, especially chlorinated hydrocarbons, were being touted as miracle products, and pesticide sales were increasing rapidly. In this climate, *Silent Spring*, parts of which also were published serially in *The New Yorker*, was at first highly controversial; one manufacturer of pesticides tried to prevent Houghton Mifflin from publishing it. Even though her own strength to publicize and to fight on the book's behalf was limited by her battle with cancer, the book received enormous attention from both the general public and policy-makers.

In *Silent Spring*, Carson's combination of literary skill and scientific knowledge enabled her to make the art of literary non-fiction a potent cultural force on specific environmental issues. By bringing together these two normally distant elements of our culture on an important public health issue, she also illustrated the power that can accompany such reconnection. Carson also, as Robert Gottlieb argues, established and defined for the public some of the major tenets of the current environmental movement. She makes the case, for example, that nature does 'not operate in closed and separate compartments', and that therefore what happens to any part of 'the web of life' (1987: 82), to the water supply or to vegetation, inevitably affects human health. Her work helped convince the American public that we humans are part of the environment, not apart from and above it.

*Silent Spring* also made the basic point, still central to the environmental movement today, that environmental concerns and especially toxicological concerns, were public issues. By always asking 'Who has decided – who has the *right* to decide – for the countless legions of people who were not consulted . . .' (1987: 127), she stresses the importance of an informed populace, the need to protect against private decisions by isolated special interests. *Silent Spring* thus established one major form of the literature of toxicity: the carefully researched, strongly supported and persuasively stated non-fictional document on a specific topic.

Of all the distinguished writers influenced by Rachel Carson, the one with the strongest acknowledged family resemblance to her is Terry Tempest Williams. Her book *Refuge* has become, in the very short period of time since its first publication in 1991, an important text for the study of environmental literature. In a 1992 Audubon essay, 'The Spirit of Rachel Carson', Williams praised her predecessor in these words:

> In 1992, I want to remember Rachel Carson's spirit. I want to be both fierce and compassionate at once. I want to carry a healthy anger inside of me and shatter the complacency that has seeped into our society.

Calling Carson's words 'a sacred text whose words became a sacrament honoring the body of the earth', Williams begins her own essay by quoting from Carson's 'A Fable for Tomorrow' section of *Silent Spring*. The quotation is significant not only because it demonstrates how Carson has shaped Williams's own views of toxicity, but also because it introduces a theme – that of an original and bountiful harmony of human and natural communities, an Eden lost by our own actions – that reappears throughout the literature of toxicity.

> There was once a town in the house of America where all life seemed to live in harmony with its surroundings. . . . Then a strange blight swept over the area and everything began to change. Some evil spell had settled on the community: mysterious maladies swept the flocks of chickens; the cattle and sheep sickened and died. Everywhere was a shadow of death. The farmers spoke of much illness among their families . . . There was a strange stillness. The birds, for example – where were they? It was a spring without voices . . . No witchcraft, no enemy actions had silenced the rebirth of new life in the stricken world. The people had done it themselves. (Williams, 1992: 104)

Along with these similarities in spirit and literary themes, Terry Tempest Williams's book *Refuge*: *An Unnatural History of Family and Place* (1991) also parallels *Silent Spring* in part of its method. Like Rachel Carson's work, Terry Tempest Williams' book gains strength from her ability to reconnect science and literature. Williams has an undergraduate degree in English and a graduate degree in science education from the University of Utah. She is both a distinguished writer and the Naturalist-in-Residence at the Utah Museum of Natural History in Salt Lake City.

As the '*Unnatural*' in her sub-title suggests, Williams uses this background to design her book along the lines of traditional natural history but with one important twist. The natural history tradition appears in her frequent and detailed descriptions of birds and habitat at the Bear River Migratory Bird Refuge on the Great Salt Lake. *Refuge*, for example, frames its narratives with an initial two page map showing the land and water forms and urban centers of the area and with a closing list, in phylogenetic order and including both common and scientific names, of the more than 200 species of birds seen there. Throughout the narrative, each chapter title consists of both a bird species and a notation of the exact water level of the lake. The *Unnatural* aspect of her title and her history refers to what humankind has done to the health of its members and the natural world, to toxic radiation and government proposals to control the natural flooding of the Great Salt Lake.

Despite all of these important similarities, however, *Refuge* is a different kind of book than *Silent Spring* and one that makes a different kind of contribution to the literature of toxicity. Instead of relying primarily on research and more general human experience, this book mixes its science – its observations from biologists, archaeologists, and ornithologists – with deeply felt, personal experience. Williams' authorial authority comes in part from her thoughtful, personal recording of her family's lives and deaths. In this highly individual family history, landscape, the landscape of Northern Utah near the Great Salt Lake, health and family are acknowledged as inextricably intertwined, and the seven generations of family interaction with the land in Utah lead, finally, to tragedy. The causes of this tragedy provoke Williams's anger at betrayal of the women in her family by the governmental patriarchy.

I should also emphasize that her emphasis also differs from the literature about nuclear weapons. It is not that the literature of nuclear power and its destructive potential on human life is new. The literature of nuclear weapons has, in fact, a long bibliography that is thoroughly discussed in Dowling's *Fictions of Nuclear Disaster* (1987). This literature is a staple of science fiction and appears in other work as well. The literature of contemporary nuclear toxicity, as represented by Terry Tempest Williams's *Refuge* in literary non-fiction and by other contemporary fiction, has, however, a different emphasis.

Terry Tempest Williams's book, which is as structurally intricate as a high modernist fictional text, moves through a combination of natural history with personal and family history by two central alternating and interlocking narratives. One narrative strand traces a seasonal natural disaster, the flooding of the Bear River Migratory Bird Refuge by the rising waters of the Great Salt Lake. This event is an especially potent one for her. From her earliest years, her grandmother (the same grandmother who talks to her about Rachel Carson) has taught her the love of landscape as a refuge. Since the age of ten, Terry has gone with her to this Bird Refuge, the nation's first waterfowl sanctuary, where the observation of its many species of birds has been a source of delight, comfort and continuity for them both. She feels that 'The birds and I share a natural history. It is a matter of rootedness, of living inside a place for so long that the mind and the imagination fuse' (p. 21). The Refuge is thus truly a refuge, and it is a place that serves for her somewhat the same purpose that the pastoral village serves in Carson's narrative. It provides a thus far uncontaminated place where the health and harmony of the landscape stand in stark contrast to the destructiveness brought about by the human created toxicity that she discovers elsewhere.

In 1983, however, the year that her narrative begins, the nearby Wasatch Mountains have their highest ever snowpack; as it melts, the runoff raises the Great Salt Lake to record high levels, covering the bird habitat in the Refuge. Parallel in time with this threat to the refuge is the renewal of her mother's long battle with cancer. Diane Dixon Tempest had recovered from breast cancer in 1971, but the disease recurs in 1983. This time it takes the form of ovarian cancer with a tumor the size of a grapefruit. In this narrative strand, time is marked by surgery, chemotherapy, radiation therapy, and resistance, a resistance which shades slowly and painfully into an acceptance of death. Hopes for victory over the disease give way slowly to an acceptance of dying. This process for Terry and for others in the family is much like a highly detailed, deeply felt, real-life version of Rachel Carson's fable of the once healthy village. A closely knit family with long ties to its region and its landscape has the existence of its female members threatened and destroyed by toxicity.

Despite a healthy diet and only one incident of breast cancer prior to 1960, the women in Williams' family are being ravaged by cancer. By the end of Williams's narrative, her mother, her grandmothers, and six aunts have all had mastectomies; seven have died, and the other two have had chemotherapy. These dead and dying women are all 'downwinders' who lived in the pattern of the prevailing winds that blew across Utah during the frequent atomic testing in Nevada in the 1950s and early 1960s. Though the government told the public that there was no cause for worry about exposure to atomic radiation, these women have all developed cancer after their exposure to toxic radiation during this nuclear testing.

The Refuge which has provided a comfortable closeness to the natural world is also threatened, but the real threat there is not the natural floods which will eventually recede, but the governmental impulse to control nature. Like the Atomic Energy Commission official who denigrated the blank spots on the map in the Utah countryside by saying: 'It's a good place to throw used razor blades', the government of Utah sees them not as '. . . an invitation to encounter the natural world, where one's character will be shaped by the landscape' (pp. 242–4), but as an empty space, a wasteland to be used. Proposals to control the flooding include nuking the Great Salt Lake (so that water would drain to the center of the earth), dying the Lake purple (to enhance evaporation), creating a giant and salt-free and thus open to development 'Lake Wasatch' from the clearwater streams which feed the lake, and the proposal finally chosen and enacted – creation of the West Desert Pumping Project. This 60 million dollar project involves an elaborate set of pumping stations, dikes and a 500 mile evaporation pond. Once it is completed, the Governor exults that 'We are finally in control' (p. 247).

Williams views these family and refuge landscapes and the environmental issues they raise from an ecofeminist perspective. Perhaps the clearest description of this important environmental philosophy and movement has been put forth by Greta Gaard in the book she edited entitled *Ecofeminism: Women, Animals, Nature* (1993): 'Ecofeminism is a theory that has emerged from various fields of feminism and activism; peace movements; labor movements; women's health care, and the anti-nuclear, environmental, and animal liberation movement.' She summarizes its central argument as the belief that 'the ideology which authorizes oppressions such as those based on race, class, gender, sexuality, physical abilities, and species' also 'sanctions the oppression of nature'. In opposition to this oppression, ecofeminism advocates '. . . a sense of self most commonly expressed by women and various other non-dominant groups – a self that is interconnected with all life' (p. 1).

The complex narrative structure of *Refuge* enables Williams to highlight her ecofeminist beliefs by juxtaposing, carefully and explicitly, abuses of the land and of women. It is, in fact, these parallels and this complexity which engage the reader, unify the book, and provide much of the emotional power for it. At first, for example, Terry's mother, Diane Dixon Tempest, is disconcerted by some aspects of her daughter's behavior. When Terry meets up with a group of red-neck hunters, men who have destroyed a long-time nest of burrowing owls in order to erect the Canadian Goose Gun Club, she walks calmly over to their pick-up truck, holds her face a few inches from the driver's face, slowly lifts her middle finger, and says: 'This is for you – from the owls and me' (p. 13).

Her mother, bound at that time by more traditional attitudes on gender, disapproves of the disrespect for men shown in this incident, one which occurs early in this text and early in her own recurrence of cancer. But, as her battles with cancer (and to some extent with the male medical establishment which is treating her) grow, she begins to question her doctor's authoritarian treatment of her body. Halfway through her struggle, she spontaneously joins other women in her family in a march for nuclear disarmament. Her changes are paralleled by similar changes in Terry's aunt who, as she fights her own cancer and medical battles, refuses to let the medical establishment take away her self-respect.

As the family narrative proceeds along this course, the governmental debates – the questions about how best to manage the rising water – move through various stages. The pumping projects discussed by the governmental fathers are built, but they become useless and ridiculous as the waters recede. Her family landscape, however, has been devastated. By the end of the two narratives, the lake has begun to recede and Terry Tempest Williams has become – at age 34 – the matriarch of her family. The book's final chapter, entitled 'The Clan of One-Breasted Women', tells how she finally discovers that her mother was not only a 'downwinder', but that she was also an eyewitness to nuclear test explosions, tests that the public had been told were harmless. Terry tells her father about a recurring dream of hers, one where she sees 'this flash of light in the night of the desert'. To her surprise, he reveals that: 'You did see it' and 'It was a common occurrence in the fifties'.

> We were driving north, past Las Vegas. It was an hour or so before dawn, when this explosion went off. We not only heard it, but felt it. I thought the oil tanker in front of us had blown up. We pulled over and suddenly, rising from the desert floor, we saw it, clearly, this golden-stemmed cloud, the mushroom. The sky seemed to vibrate with an eerie pink glow. Within a few minutes, a light ash was raining on the car. (p. 283)

Despite the repeated governmental assurances that no harm would be done to those people exposed, the cancers of the Williams family women have appeared fourteen years after their exposure. Fourteen years, she notes, is the time that 'Howard L. Andrews, an authority in radioactive fallout at the National Institutes of Health, says radiation cancer requires to become evident' (p. 286).

Terry Tempest Williams's contribution to the literature of toxicity moves toward conclusion with a powerful dream sequence that brings her public behavior together with her personal beliefs. As she and several other women stage a dance of protest against nuclear testing, her dream then shades into a real life protest, an act of 'civil disobedience' against the conducting of nuclear tests in the desert. When an officer detains her and questions the purpose of a pad of paper and a pen found in her boot top, she replies that they are 'weapons'. Terry Tempest Williams' personal narrative in *Refuge* thus makes a point much in harmony with Rachel Carson's assertion, near the end of *Silent Spring*: 'The "control of nature" is a phrase conceived in arrogance, born of the Neanderthal age of biology and philosophy, when it was supposed that nature exists for the convenience of man' (p. 297). Despite the differences in time of publication and in literary form, the two books share elements of an ecofeminist perspective on toxicity.

Aside from Rachel Carson's *Silent Spring* and Terry Tempest Williams's *Refuge*, the majority of influential contemporary literary treatments of toxicity occur in fiction. Two of the most prominent novels to develop this theme are Jane Smiley's *A Thousand Acres* (1991), a book which was awarded both the National Book Critics Circle Award and the Pulitzer Prize, and Don DeLillo's *White Noise* (1985), a book which is frequently taught in university classes on contemporary fiction and which won the National Book Award. Since these books develop their common theme of toxicity in very different ways, they serve to illustrate the diversity in the treatment of this theme as well as its prominence.

Smiley's book, like Williams's *Refuge*, is a compelling work of ecofeminism. Smiley sets her novel on a huge midwestern family farm in the year 1979, just before such family farms begin falling into foreclosure. As she revealed in an interview with Charles Hix (1991), however, her goal was more than to talk about 'farming as practiced on big industrial farms in America today'. Her interest was also in feminist theory and in abusively dysfunctional families, and Smiley's complex embodiment of ecofeminist perspectives enables her to unite all these elements in a major novel.

Like Carson and Williams, Smiley's depiction of the contemporary world unfolds within the context of an earlier time of ecological health and natural abundance. The narrator, Ginny Smith, opens her story with a walk along the Zebulon River that evokes the fertility of the wetlands before the settlers drained them for farming. 'And there was a flock of pelicans, maybe twenty-five birds, cloud white against the shine of the water. Ninety years ago when my great-grandparents settled in Zebulon County and the whole county was wet, marshy, glistening like this, hundreds of thousands of pelicans nested in the cattails, but I hadn't seen even one since the early sixties' (p. 9).

This decline in waterfowl is matched by the loss of human intimacy with the land. What the absence of wetlands means for the women of her family is made clear in Ginny's thoughts on a pond which used to be on their property. Once the pond, 'an ancient pothole that predated the farm', was used by the sisters as a place to swim. She relates that their Daddy, taking action, significantly, at a time 'not long before the death of our mother', drained this pond so he could work that field more efficiently. Then, as all the little lakes and pothole ponds like this one disappear, the city attempts to replace them by building a swimming pool. The power of a concrete pool to replace the natural landscape is severely limited, however. As Ginny takes her sister Rose's two daughters to the town pool, she remembers and mourns. 'What Rose and I once did in our pond, simply float on our backs for what seemed like hours, soaking up the coolness of the water and living in the blue of the sky was impossible here. There was no place to be privately, contemplatively immersed' (p. 95).

This loss of intimacy with nature is explicitly tied to gender. Not only is it the women who mourn the loss of intimacy with nature, it is the men, as they work to build the huge family farms, who have drained the swampy ground and destroyed the wetlands. The gender differences implied in these attitudes toward the land are underlined when Ginny relates the history of this drainage process. Ginny's father, Laurence Cook, who has built the one thousand acre Cook farm, represents the attitude of the men who have labored for so many years to lay the drainage tile. He always speaks of the early, unaltered wetlands with distaste, and he is so physically imposing and so self-righteous, his domination of the landscape so total, that Ginny can never see him in perspective. Even in relation to the wide sweep of prairie land, 'He was never dwarfed by the landscape . . .' (p. 20). The men of the family adopt an arrogant, God-like attitude toward the land and nature, an attitude that Ginny summarizes in this way: 'However much these acres looked like a gift of nature, or of God, they were not. We went to church to pay our respects, not to give thanks' (p. 15).

The men's allies in this effort are the farm machines and the chemical pesticides. Their real love, the mature Ginny who narrates the novel observes, is for 'new, more efficient equipment'. With these machines, they create acres of 'monochromatic greenness' and corn which grows with 'mechanical uniformity' (p. 152). With them, they also continue the process of wildlife eradication that the wetland drainage has begun, killing deer and rabbits and birds in their relentless plowing of the fields for more efficient crop production.

This domination of the land extends also to extensive pesticide use, the consequences of which strike the women as well as the land. Early in the novel, we learn that Ginny has had five miscarriages. As Ginny recites the family history that makes up much of the novel's narrative, she gradually reveals that her mother died of cancer at a very young age, that the mother of a neighboring family, the Clarks, also died of cancer, and that her sister, Rose, first has breast cancer and a radical mastectomy and then dies of cancer while still very young.

Part of the power of Smiley's narrative is that we see Ginny's slow and painful progress in recognizing the truth about this situation. She, like the rest of the Cook family and most of the other farmers in the area, is at first totally immersed in the prevailing farm culture. For a long while, she does not make the crucial connection between the toxicity of the pesticide seeping into the acquifer and the prevalence of cancer in the women. It is only when a partial outsider, a neighbor's son, returns after 13 years away and states the connection directly to Ginny that she finally starts to see how her string of miscarriages is linked to the abuse of nature. By the end of her narrative, Ginny finally can acknowledge that part of her heritage is 'the loop of poison we drank from, the water running down through the soil, into the drainage wells, into the lightless mysterious underground chemical sea, then being drawn up, cold and appetizing into Rose's faucet, my faucet' (p. 370).

Another part of Ginny's family legacy furthers this ecofeminist connection between abuse of the land and abuse of women, making Smiley's point that the patriarchial culture treats them both as things, objects to be dominated and used, rather than respected. The day of their mother's funeral, their Daddy takes over a neighbor's farm by unscrupulous means. With the mother gone, the father then bullies Ginny into sex with him and also seduces Rose, her tougher, less compliant sister. Ginny, however, has kept this horrible knowledge buried, tucked away from her own adult awareness. So traumatic is her father's sexual abuse of her that, although fear and shame are the dominant emotions she experiences as an adult, the actual specific memory has been unconsciously lost. It is only as she starts to relive her childhood in the face of her father's increasing senility and her sister Rose's relentless probing that Ginny comes to remember, first, her father's insistence on always having his own way, then his beatings of her and her mother's refusal to help, and finally, his sexual abuse of her. She gradually comes to acknowledge the truth of Rose's assertion that 'We were just his, to do with as he pleased, like the pond or the houses, or the hogs or the crops' (p. 191).

This anger at the father who beat them, abused them and still retained the respect of the men in his community finally forces Ginny to go beyond mere agreement with Rose's viewpoint and to make her own judgments. When Ginny thus finally describes Rose's cancer as 'her dark child, the child of her union with Daddy' (p. 323), the connection she makes through this metaphor is a critical one. Her recognition that the incestous abuse of the sisters is tied to the chemical abuse of the land and thus to the cancer caused by it is, in many ways, the climatic event in *A Thousand Acres* – a crucial concluding insight in a novel detailing the education of an ecofeminist. Like the women in Terry Tempest Williams's family history, Ginny Smith tries to find ways to trust the patriarchial management of land and society and thus to avoid conflict, but her experience has lead her to adopt, albeit somewhat reluctantly, a feminist viewpoint.

Don DeLillo's highly stylized *White Noise* (1985) embodies concern with toxicity in a very different way. Instead of a deliberately paced narrative chronicle of one individual's slow awakening to a pervasive toxicity in gender and land relationships, DeLillo builds his fictional world with a fast moving mix of satire and deadly seriousness. He describes an American landscape saturated with toxicity, and he creates an American family obsessed with this toxicity and its consequences.

His novel, following the conventions of the emerging literature of toxicity, considers and quickly rejects the possibility of an idyllic, pastoral, healthy village life. The narrator, Jack Gladney, chairs the department of Hitler Studies at a generically named 'College-on-the Hill' where 'tuition is fourteen hundred dollars, Sunday brunch included' (p. 41). He lives with his current wife, Babette, in the college town of Blacksmith 'at the end of a quiet street in what was once a wooded area with deep ravines' (p. 4). Although this village is far from a major city and the villagers would like to be free from urban 'history and its contaminations,' they cannot escape. Even here, 'Dying is a quality of the air' (p. 38). Family and friends are preoccupied with various forms of toxicity that range from the 'white noise' generated by supermarket door openers, television waves, radio transmissions and appliances to the troops of men, dressed in Mylex suits, respirator masks and boots, who appear at crucial junctures in the narrative to handle simulated disaster drills or actual toxic clean-up.

DeLillo's sharp satire of contemporary American life also introduces a new and important characteristic of the literature of toxicity – the link between it and unchecked American materialism, a preoccupation with an abundance of things. The passage of seasons on this village campus are, for example, not marked by natural changes, but instead by a huge migration of consumer goods. Jack's mock-epic catalogue of the students' return to campus is both humorous and tragic:

> The roofs of the station wagons were loaded down with carefully secured suitcases full of light and heavy clothing: with boxes of blankets, boots, and shoes, stationery and books, sheets, pillows, quilts, with rolled rugs and sleeping bags, with bicycles, skis, ruck sacks, English and Western saddles, inflated rafts. As cars slowed to a crawl and stopped, students sprang out and raced to the rear doors to begin removing the objects inside: the stereo

sets, radios, personal computers, small refrigerators and table ranges; the cartons of phonographic records and cassettes; the hairdryers and styling irons; the tennis rackets, soccer balls, hockey and lacrosse sticks, bows and arrows; the controlled substances, the birth control pills and devices; the junk food still in bags – onion and garlic chips, nacho thins, peanut creme patties, Waffelos and Kabooms, fruit chews and toffee popcorn, the Dum-Dum pops, the Mystic mints. (p. 3)

The first pages of this novel emphasize how the constant 'daily seeping falsehearted death' of toxicity has compromised the future of their children. Does, for example, the receding hairline of Jack's 14-year-old son Heinrich tie to 'some kind of gene-piercing substance consumed by his pregnant mother', or, perhaps, to an unwitting location 'in the vicinity of a chemical dump site in the path of air currents that carry industrial wastes capable of producing scalp degeneration. . . .?' (p. 22). Wilder, the young son of Jack and his current wife Babette, acts even younger than his years, and his vocabulary growth has stopped at 25 words. He goes on a mysterious crying jag that lasts for seven hours. When the local grade school must be evacuated because the kids are 'getting headaches and eye irritation, tasting metal in their mouths' (p. 35), their problems could be from any of a multitude of sources – the ventilation system, foam or electrical insulation, or asbestos particles.

In 'The Airborne Toxic Event', the novel's second section, the focus falls on the adult generation and specifically on the immediate consequences of a train tank car derailment in Blacksmith that releases 'a heavy black mass. . . .' (p. 110). This 'shapeless growing thing' (p. 111) results from the spill of a whole, new generation of toxic waste by-products generated in the manufacture of a pesticide called Nyodene Derivative or Nyodene D. The government bureaucracy first tries to contain the disaster verbally, but their attempts to provide distance and detachment by designating it, in typically bureaucratic language, as an 'airborne toxic event' fail to allay the villagers' fear.

The rumored reports of Nyodene D's effects on humans escalate rapidly; the exposure symptoms attributed to the drug move from skin irritation and sweaty palms, to nausea, vomiting, and shortness of breath, and finally on to convulsions, coma and miscarriage. It slowly emerges that specialists in such matters know that this toxic chemical grows lumps in mice, that it has a life span of 40 years in the soil and of 30 years in humans. Heinrich, who is precocious in his knowledge of toxic chemicals, explains that 'no one seems to know exactly what it causes in humans or in the offspring of humans. They tested for years and either they don't know for sure or they know and aren't saying. Some things are too awful to publicize' (p. 131).

The toxic spill and cloud lead to the 'decorous hysteria' of a full-scale evacuation, presided over by the men in Mylex suits and by vehicles which move through the streets broadcasting something which sounds like 'cloud of deadly chemicals, cloud of deadly chemicals'. Although this mass movement of people has an epic and far-reaching quality to it as it is 'connected in doom and ruin to a whole history of people trekking across wasted landscapes' (p. 122), it is Jack Gladney who is most exposed. He gets out to fill his car with gas just before the black cloud, accompanied by seven helicopters with spotlights on it, passes overhead. His view of it as it 'moved like some death ship in a Norse legend escorted across the night by armored creatures with spiral wings' (p. 127) is prophetic. Soon afterwards, Jack has himself tested for toxicity and tells his friend and colleague Murray: 'The little breath of Nyodene has planted a death in my body' (p. 150).

In the final section of *White Noise*, Jack's apprehension of his death sentence grows stronger. He goes for further tests to a medical facility called Autumn Hill Farms where he learns that he has 'a nebulous mass' in his body, a growth which comes from and corresponds to the original 'heavy black mass' of toxic cloud. He and Murray discuss possible ways to deal with this death sentence, including trying to tap what Murray calls 'A great dark lake of male rage' (p. 292) as a way to beat death. (Space precludes exploring here the ties between Murray's theory of male anger and the ecofeminist indictment of environmental abuses in patriarchial culture.) None of these works, however, and Jack's only answer is to become preoccupied with throwing away things. In a 'vengeful and near savage state', bearing 'a personal grudge against these things', he throws away everything from hip boots to diplomas, from padded clothes hangers to magnetic memo clipboards, feeling that things somehow had 'put me in this fix' (p. 294). He even rejects that ultimate consumer device, that seemingly magical multiplier of things – a new ATM card. The novel, however, suggests that these gestures come too late. In one of the final views of Jack, he is standing with his family and other village residents on an overpass watching one of the especially beautiful sunsets which have appeared often since the toxic event.

This elegiac endnote suggests the end of a civilization as well as a single man. Despite the flashes of brilliant humor in DeLillo's depiction of contemporary life, he is more pessimistic than either Williams or Smiley. Their depictions of toxicity are somewhat more singular and localized than his creation of a country saturated with multiple and omnipresent toxicities. Unlike Terry Tempest Williams whose recognition of what has been done to her and those she loves leads to civil disobedience and thus to possible social change, and unlike the character of Ginny Smith whose moment of recognition leads to anger and to her move away from the toxic farm landscape, Jack Gladney's story concludes with a sense of resignation.

## III

Early in part two of *White Noise*. Jack Gladney says: 'I don't see myself fleeing an airborne toxic event. That's for people who live in mobile homes out in the scrubby parts of the country, where the fish hatcheries are' (p. 117). This outlandish statement – DeLillo's satiric commentary on the uneasy mix of complacency and defensiveness that he sees as characterizing the American middle class's feelings about toxicity – does not stand alone. Gladney elsewhere tries to relegate toxic health threats only to 'the poor people who live in exposed areas' and 'the poor people and the uneducated who suffer the main impact of natural and man-made disasters' (p. 114).

Gladney's notions of a privilege fail, of course, to protect him. In the world of DeLillo's novel, toxicity lurks everywhere and the traditional refuge of membership in a protected social class is actually no refuge at all. It does, however, point us toward what has been a kind of truism in the study of environmental literature: those social groups whose homes and jobs are over-represented in their proximity to toxicity are the same ones who have been somewhat under-represented in the literature about nature and the environment (see Schwab, 1994). Important recent works by Rudolfo Anaya and Ana Castillo indicate that the literary part of this equation is changing, that the social issues of class and ethnicity are beginning to figure more prominently in environmental literature.

Rudolfo Anaya's powerful short story, 'Devil Deer' (1993), is a crucial example of this development. To fully understand this story's power and its important place in the unfolding history of environmental literature, we need first to look briefly at the author's lifelong concern with our human ties to the natural world. Raised in the southwest, in Santa Rosa, New Mexico, Anaya clearly sees himself as shaped by its landscapes: 'I don't believe a person can be born and raised in the southwest and not be affected by the land. The landscape changes the man, and the man becomes his landscape. My earliest memories were molded by the forces in my landscape: sun, wind, rain, the llano, the river.' Nature, he argues then, is a source of strength:

> In speaking about landscape, I would prefer to use the Spanish word *la tierra*, simply because it conveys a deeper relationship between man and his place, and it is this kinship to the environment which creates the metaphor and the epiphany in landscape. On one pole of the metaphor stands man, on the other is the raw, majestic and awe-inspiring landscape of the southwest: the epiphany is the natural response to that landscape, a coming together of these two forces. And because I feel a close kinship with my environment I feel constantly in touch with that epiphany which opens me up to receive the power in my landscape. (1977: 98–9)

These views shape Anaya's early writing and are most notably embodied in his classic novel, *Bless Me, Ultima* (1972); in it, the young man Antonio, aided by the healer Ultima, is led to see the positive energy latent in the landscape. A powerful, life-affirming and life-shaping epiphany much like that described above in Anaya's non-fictional prose becomes a turning point in Antonio's life. In the short story 'Devil Deer' published some 21 years later, this joyous, epiphanic sense of beneficent connectedness gives way to a much darker vision. Toxicity and its power to disfigure beauty and to destroy life have become the focus.

In the beginning of this story, the protagonist Cruz and other members of his Pueblo plan, with stories and celebration, for a subsistence deer hunt in the Jemez Mountains. In this story's version of the now familiar theme of a toxic-free community living in harmony with each other and with nature, the corn and pepper harvests have already been brought into the pueblo, and the progression of the seasons has brought the village to deer season. Autumn has arrived, 'clean and sharp and well-defined', and the 'falling leaves of the aspen were showers of gold coins'. This is a time when the women joke with the men about their hunting prowess and look ahead to the deer meat that will provide winter sustenance for the family. It is a time as well when the men plan the hunt with their friends, anticipating 'a ritual shared since immemorial time' (p. 112–13).

But, in the midst of this happiness, there is also a bleaker note. 'Everybody knew the deer population was growing scant. It was harder and harder to get a buck. Too many hunters, maybe. Over the years there were fewer bucks. You had to go deeper into the forest, higher, maybe find new places, maybe have strong medicine.' For this reason, Cruz has persuaded his friend Joe to go with him to a place called Black Ridge. Since part of this ridge is fenced in by the Los Alamos Laboratory, few people hunt there. None the less, when Joe breaks his leg, Cruz – despite his friend's warning and his own sense that 'the ridge lay silent and ominous on the side of the mountain' (p. 114) – resolves to hunt alone.

The initial signs are not encouraging. Cruz falls asleep in his truck, clutching the leather bag that holds his stone black bear fetish and dreaming a nightmare in which a black bear speaks to him:

> The bear was deformed. One paw was twisted like an old tree root, the other was missing. The legs were gnarled, and the huge animal walked like an old man with arthritis. The face was deformed, the mouth dripping with saliva. Only the eyes were clear as it looked at Cruz. Go away, it said, go away from this place. Not even the medicine of your grandfathers can help you here. (p. 115)

When he wakes, Cruz resolves to stay and get his deer, despite the 'eerie, blue glow' which fills the night and which suggests to him that the mountain was dying. He stalks a buck, who seems also, mysteriously, to be stalking him, then follows the buck through a tear in the fence and shoots him. As he completes the hunt and kills his 'brother' the deer, Cruz asks himself what causes the vibration beneath his feet. 'What kind of devil machines were they running over in the labs that made the earth tremble? Accelerators. Plutonium. Atom smashers.' For the first time ever, he was afraid on the mountain.

When he approaches the dead deer, the bad omens are fulfilled.

> The deer was deformed. The hide was torn and bleeding in places, and a green bile seeped from the holes the bullets had made. The hair on the antlers looked like mangy, human hair, and the eyes were two white stones with mottled blood. The buck was blind. Its legs were bent and gnarled. That's why it didn't bound away. The tail was long, like a donkey tail. (p. 119)

Like Ginny Smith in *A Thousand Acres*, Cruz comes reluctantly to a moment of horrible recognition. The devil deer forces him, finally, to the conscious realization of how bad things really are, to see what he has tried to deny – that the toxicity of the nuclear plant is a threat to the healthy life of his people. 'Whatever it was, it was seeping into the earth, seeping into the animals of the forest. To live within the fence was deadly, and now there were holes in the fence' (p. 119).

With this, Cruz faces the reality that comes at one time or another to all of the central characters in the literature of toxicity since Rachel Carson. Despite humankind's ingenuity in seeking protective barriers and despite all of the government's promises of safety, toxicity, once created, cannot be contained. When Cruz takes the deer back to show the old men of the pueblo, his 'voice and vacant stare' reveal the enormous sadness of the lesson. This initation is the precise opposite of that experienced by Antonio in *Bless Me, Ultima.* Instead of awakening to the landscape as a source of power, Cruz recognizes that toxicity has turned his landscape into a destructive force. The medicine men will perform a cleansing ceremony, but Anaya's short story closes with an unanswered worry: 'Did they have enough good medicine to wash away the evil the young man had touched?' (p. 121). The question measures how Anaya's concern about toxicity has led him far from his early optimism about the relationship between humans and the land.

Like Rudolfo Anaya, Ana Castillo makes environmental concerns central to her fiction. The larger canvas available in her novel *So Far From God* (1993), however, also enables her to embody a more sweeping social indictment of environmental victimization. Castillo's longer work tells the sad story of Fe, a central character destroyed by her work with toxic chemicals, but it also speaks, more explicitly and broadly than any work considered thus far, for those whom Jack Gladney describes as 'the poor people who live in exposed areas', the socio-economic and ethnic groups living on the margins of American affluence and power.

Castillo's novel about the lives and deaths of the four daughters of Sofi, La Mayor of the Village Council of the village of Tome in central New Mexico, deals with environmental questions throughout its many wide-ranging and interwoven stories. From an opening dedication 'To all the trees that gave their life to the telling of these stories', through concerns expressed about such things as 'noise pollution coming from nearby Kirtland Air Force Base' (p. 225), to a brilliant concluding scene which remakes the traditional Way of the Cross

Procession into a catalogue of toxic ills visited upon the poor, Castillo continually comes back to environmental issues.

The novel's central environmental narrative, however, is the story of Sofi's daughter Fe, and Castillo begins this narrative with a evocation of a healthy community living in harmony amidst nature's bounty.

> It was that month in the Land of Enchantment when it smelled of roasted chiles everywhere. Fresh red ristas and sometimes green ones were hung on the vigas of the portales throughout, all along dusty roads, in front of shops and restaurants to welcomes visitors and to ward off enemies. (p. 170)

This apparent harmony, however, is only superficial. Fe chooses this same harvest month as the time for her marriage, an event that fuels her pursuit of a materialism akin to that depicted in DeLillo's novel, the purchase of 'the long-dreamed-of automatic dishwasher, microwave, Cuisinart, and the VCR' (p. 170). Fe's efforts to 'have a life like people do on TV' is made possible by the money earned at her new job and 'it was the job that killed her' at the age of 27, exactly one year later.

The job is with a big new company, Acme International, where Fe cleans parts for high-tech weapons assembly on a government contract. With wages and bonuses, the job offers twice the pay of her old one. The problem is that the women employees get nausea and migraines that increase day by day. When they talk to the company nurse about these symptoms, she gives them 'ibuprofen tablets, advice about pre-menopause and the dropping of estrogen levels in women over thirty, and pretty much tells them that their discomfort was just about being a woman and had nothing to do with working with chemicals' (p. 178).

Since Fe has her high school diploma, speaks English as her first language, and works with dedicated efficiency, she wins pay bonuses and 'special' assignments. Despite her miscarriage and the eerie coincidence of many such cases at the plant, her strong faith in the system (emphasized by Castillo's choice of a name that means faith in Spanish) leads her to continue working hard for Acme. After three months of using a chemical that actually glows in the dark, she becomes tired and irritable; her breath smells like glue. Listerine, however, 'was not going to wash out what her lungs and liver and kidneys had already absorbed' (p. 182).

Fe's next reward is an even tougher job, one where she is given an unlabelled chemical and sent to work in isolation in the basement. Although the chemical eats up her orange gloves and dissolves her fingernails, the company provides no vent and no mask. She pours the left-over chemicals down the drain until an angry supervisor tells her that she must let it evaporate from the pan. Despite these warning signs, Fe's loyalty and efficiency keep winning raises for her until the point that she has 'big dried spots on her legs' and 'one constant fire drill going on in her head' (p. 185). Her mother's insistence on a medical check-up leads to the discovery that Fe has cancer.

After undergoing months of torturous medical treatment, misplaced catheters which feed chemotherapy supplies to the wrong part of her body, and accusations by the FBI that she deliberately used illegal chemicals, Fe returns briefly to the plant so she can insist on reading the manual for the chemical she used. It is only then that she learns that this toxic cleaner is 'heavier than air', that it must not be left to evaporate. Fe then screams out her recognition that the deadly chemical has gone into her body.

Like Cruz and like Ginny Smith, Fe comes reluctantly to face the fact that toxicity threatens her existence; like Terry Tempest Williams, she has been betrayed by those in whom she had faith. Before her death, Fe also has a moment of recognition much like Ginny's awakening in *A Thousand Acres*, one where she realizes that the larger circle of community life has been poisoned as well, that the chemical she 'dumped down the drain at the end of the day' permeates the community water system as it goes 'into the sewage system and worked its way to people's septic tanks, vegetable gardens, kitchen taps and sun-made tea' (p. 188). The poison cannot, as Jack Gladney tried to convince himself, be relegated to any one person or group; it affects everyone.

Castillo also goes back to the beginnings of the literature of toxicity in a description reminiscent of Rachel Carson's village in *Silent Spring*, the village after it has been transformed by toxic pesticides. These echoes are of Carson's village after 'a strange blight crept over the area', when 'mysterious maladies swept the flocks of chickens; the cattle and sheep sickened and died'. Children and adults die unexpectedly, and Rachel Carson symbolizes all this destruction in a strange stillness – a silent birdless spring. Castillo's view of the consequences of toxic degradation, set some thirty years later in the village of Tome, New Mexico, is notably similar.

> And meanwhile, most of the people that surrounded Fe didn't understand what was slowly killing them, too, or didn't want to think about it, or if they did, didn't know what to do about it anyway and went on like that, despite dead cows in the pasture, or sick sheep, and that one week late in winter when people woke up each morning to find it raining starlings. Little birds dropped dead in mid-flight, hitting like Superball hail on roofs, collecting in yards and streets and falling on your head if you didn't look out. Unlike their abuelos and visabuelos who thought that although life was hard in the 'Land of Enchantment' it had its reward, the reality was that everyone was now caught in what had become: The Land of Entrapment. (p. 172)

In its view of nature and society threatened by toxicity, *So Far From God* is, sadly, not so far from *Silent Spring*. Thirty-one years later, the fears are unabated.

Castillo does differ from earlier writers in the literature of toxicity, however, in that she concludes her presentation of environmental toxicity with a dramatic reminder of environmental inequalities. Toxicity is a threat everywhere, allowing no exemptions, but she does raise her voice most strongly on behalf of those who are too often voiceless, those who often remain less important to governmental and business and even environmental interests. As Castillo moves toward the end of *So Far From God*, she features a march which, like the march at the end of *Refuge*, features women speaking out on environmental ills. She, however, uses the framework of a *penitentes* march, instead of a protest march, and this Way of the Cross Procession is unlike anything anyone had ever seen. It has no brothers flagellating themselves with horsehair whips, and no brother elected to carry 'a life-size cross on his naked back'. The theme throughout emphasizes how toxic abuse is 'turning the people of those lands into an endangered species' and each station of the cross catalogues a different threat. These range from dumping radioactive waste in the sewers and in contaminated canals where children and livestock drink and die from it, to air pollutants coming from the factories, and to 'deadly pesticides' sprayed directly from the helicopters onto 'the vegetables and fruits and on the people who picked them for large ranchers at subsistence wages' (p. 241–5).

A woman with a baby in her arms addresses those who, like Jack Gladney, assume an attitude of callous disregard towards her and those like her. Her short speech effectively summarizes this most recent development in the literature of toxicity.

> We hear about what environmentalists care about out there. We live on dry land but we care about saving the whales and the rain forests, too. Of course we do. Our people have always known about the interconnectedness of things; and the responsibility we have to 'Our Mother,' and to seven generations after our own. But we, as a people, are being eliminated from the ecosystem, too . . . like the dolphins, like the eagle; and we are trying to save ourselves before it's too late. Don't anybody care about that? (p. 242)

## IV

Historically, American prose nature writing has been seen as almost exclusively literary non-fiction; those who write about nature often have been referred to as the 'sons and daughters of Thoreau'. In the last third of the twentieth century, however, as the creation and manufacture of toxic materials have increased and our literary efforts to oppose environmental degradation are broadening, multiplying and diversifying, it is not surprising that we now have an important group of writers who fall into a grouping which should be called 'the daughters and sons of Rachel Carson'.

Although Carson's original efforts to bring science and literature together in literary non-fiction have not been sustained at her high level, both the quality of these literary works and the importance of the prizes awarded to them underline a significant part of the larger Carson legacy: the assurance that prose dealing directly with society's toxic problems can also be important literature. The sadness that attends Carson's bequest is that the need for such literature is now as great, perhaps even greater, as when *Silent Spring* first appeared. We do not yet have enough indications that, in Rudolfo Anaya's terms, the good medicine has begun in earnest to wash away the threats depicted in the literature of toxicity.

*AUTHOR'S NOTE*

The origins and the range of the literature of toxicity can bear much more discussion than this chapter has had space to include. Some students of literature might argue its origins in works of English fiction as early as the urban novels of Charles Dickens. Those involved in the history of medicine might argue that the literature rightly begins with Alice Hamilton's classic 1925 study *Industrial Poisons in the United States*. Since space does not permit a comprehensive study, I have chosen to focus on the contemporary literature of toxicity which does clearly date from Rachel Carson's 1962 book. I should, however, also emphasize some other choices that structured this chapter. I did not attempt a thoroughgoing survey of all the contemporary works which would need to be included in a comprehensive study. There has not been room to talk, for example, about important poems which develop this theme, nor even to mention all the works of literary non-fiction and fiction which have followed Rachel Carson's *Silent Spring* on the contemporary scene. Walker Percy's *The Thanatos Syndrome*, to name only one example, deserves analysis and discussion in and for itself and for the light which almost surely would come from the juxtaposition of it with DeLillo's *White Noise*. The effort here has been, instead, to present representative contemporary works, works which both illustrate the high quality of literary creativity at work on this theme and which represent, as well, the important currents as the theme develops. It is also clear that each of the texts studied does deserve more time and consideration than could be given here, but my hope is that bringing together this material will help foster other scholarship on what is surely one of the most critical topics in environmental studies.

## REFERENCES

Anaya, Rudolfo (1972) *Bless Me, Ultima*, Warner Books–Time Warner, New York.

Anaya, Rudolfo (1977) The writer's landscape: epiphany in landscape. *Latin American Literary Review*, V(1) (Spring-Summer), 98–102.

Anaya, Rudolfo (1993) Devil Deer, in *Southwest Stories* (ed. John Miller), Chronicle Books, San Francisco, pp.112–21.

Brooks, Paul (1989) *The House of Life: Rachel Carson at Work*, Houghton Mifflin, Boston, MA.

Carson, Rachel (1981) *The Sea Around Us*, Oxford University Press, New York.

Carson, Rachel (1987) *Silent Spring*, Houghton Mifflin, New York (first published 1962).

Castillo, Ana (1993) *So Far From God*, Plume, New York.

DeLillo, Don (1985) *White Noise*, Penguin, New York.

Dowling, David (1987) *Fictions of Nuclear Disaster*, University of Iowa Press, Iowa City.

Gaard, Greta (1993) Living interconnections with animals and nature, in *Ecofeminism: Women, Animals, Nature* (ed. Greta Gaard), Temple University Press, Philadelphia, pp. 1–12.

Glotfelty, Cheryll Burgess (1989) Toward an Ecological Criticism. Paper given at a Western American Literature Meeting, Coeur D'Alene, Idaho, 13 October 1989.

Gottlieb, Robert (1993) *Forcing the Spring: the Transformation of the American Environmental Movement*, Island Press, Washington, DC, pp. 81–6.

Hix, Charles (1991) Interview with Jane Smiley. *Publishers Weekly*, **8**, 20.

Kroeber, Karl (1993) *Ecological Literary Criticism: Romantic Imagining and the Biology of Mind*, Columbia University Press, New York.

Leuders, Edward (1989) *Writing Natural History: Dialogues with Authors*, University of Utah Press, Salt Lake City.

Love, Glen A. (1990) Revaluing nature: toward an ecological literary criticism. *Western American Literature*, **25**(3) 201–15.

Meeker, Joseph (1974) *The Comedy of Survival: Studies in Literary Ecology*, Charles Scribner's Sons, New York.

Schwab, Jim (1994) *Deeper Shades of Green: the Rise of Blue-Collar and Minority Environmentalism in America*, Sierra Club, San Francisco.

Smiley, Jane. (1991) *A Thousand Acres*, Ballantine Books, New York.

Snow, C. P. (1963) *The Two Cultures and the Scientific Revolution*, Cambridge University Press, New York.

Williams, Terry Tempest (1991) *Refuge: An Unnatural History of Family and Place*, Vintage Books–Random House, New York.

Williams, Terry Tempest (1992) The spirit of Rachel Carson. *Audubon*, (July–August), 104–7.

# *Synthesis*

# 17 *Ecosystem degradation: links to human health*

EMILY MONOSSON AND RICHARD T. DI GIULIO

When we first described the nature of this book to potential authors, one ecologist we spoke with emphasized the need for a synthesis chapter. He said 'There are so many books out there that are just a collection of chapters – with no synthesis – isn't that the goal of this sort of book?' Our goal was to get the lines of communication open, relating to the topic of human health and well-being and that of the environment, across several different fields of inquiry. Common themes emerged throughout this book, illustrating linkages between vastly different fields. The goal of this chapter is to identify the common threads carried through the chapters of this book.

While putting the book together the most common question, from molecular biologists, ecologists or literary scholars, was 'What do you mean by ecosystem health?' Human health has some meaning to everyone, but we could not define what we meant by ecosystem health. In fact we considered dropping it from the title, but could not think of an appropriate term or phrase to replace it! Instead, we left it up to each author to use the term or not, according to their own understanding of the term. As discussed in Chapter 1 of this book, the term is quite controversial even with, or perhaps because of, its lack of definition.

Regardless of the 'fuzziness' of the term, it is useful in getting the message across. What we are really talking about is not necessarily 'ecosystem health', but perhaps lack of it: ecosystem degradation or ecosystem change. Effects on human health can be linked to ecosystem degradation as discussed in many of these chapters, from the adverse effects of environmental contaminants on wildlife individuals or populations to the reduced or altered function of an ecosystem. Lack of definition aside, there probably would be no such term if

*Interconnections Between Human and Ecosystem Health.* Edited by Richard T. Di Giulio and Emily Monosson. Published in 1996 by Chapman & Hall, London. ISBN 0 412 62400 1.

humans did not have such a great impact on their environment, nor would the topic for this book exist.

The subjects covered by the first chapters concern interconnections at the most basic levels of biological organization – the subcellular or molecular levels, followed by chapters about interconnections at the ecosystem or landscape level. These contributions set the stage for the second half of this book which addresses sociological and economic issues involved in defining ecosystem health for management purposes. By the end of these chapters we begin to understand the importance of society's values for the application of effective ecosystem management. The final two chapters attempt to understand society's relationship with the environment, which in turn influences its values, through literary analysis. As one proceeds through the book, the relationships between humans and the environment that serve as the basis for each chapter become more complex and multidimensional. The authors discuss interconnections they feel are important in their own fields. Surprisingly there are many similarities despite the diversity of disciplines.

We begin our synthesis at the subcellular level, the smallest level of biological organization discussed in this book. Here the interconnections are most clear, perhaps because the components are easily defined (DNA, receptors, steroid hormones) and their role in the 'health' of an organism is fairly clear, although even at this level there are many unknowns. Chapters 2–4 fall into this category: a discussion of the molecular mechanism of tumorigenesis by Van Beneden, the phylogeny of the Ah receptor by Hahn and the effects of various endocrine-disrupting chemicals on the reproductive system by Gray *et al.* All three chapters describe highly conserved systems that are present in all vertebrates tested, and some of which also occur in invertebrates, such as the presence of certain oncogenes discussed by Van Beneden. Each author discusses environmental contaminants that can interact with these systems resulting in adverse effects. The question of interconnection is addressed at the level of individual organisms or populations within an ecosystem. Humans share the same conserved systems with a great number of wildlife species. Chemicals that bind the Ah receptor, turn on oncogenes or disrupt the endocrine system are capable of affecting both the inhabitants of polluted ecosystems and humans as well. Ecosystem health at this level refers to the health of the individuals within an affected ecosystem.

These chapters raise the question – where do humans fit in? What are the boundaries of a particular ecosystem? Do they include human populations? Are humans as susceptible to the effects of environmental contaminants as some wildlife species? These questions are addressed at different levels in many of the remaining chapters. In the next two chapters, by Colborn and Adams *et al.* (Chapters 5 and 6 respectively), humans are linked directly with affected ecosystems – and as a result we are exposed to the same contaminants as are many wildlife species. Both chapters illustrate quite clearly that wildlife species can be affected by environmental contaminants, and many of the documented wildlife effects discussed in these chapters suggest exposure to endocrine disrupters, or reproductive toxicants.

It is often difficult to detect effects of exposure to environmental contaminants in human populations. There are many different routes of exposure depending on the type of contaminant and the matrix (i.e. soil, water, air, food). In some cases, adverse effects observed in human populations can be linked to chemical exposure (e.g. mercury poisoning in Japan, lead poisoning in children in urban areas). Chemicals that cause cancer or disrupt the endocrine or reproductive system, however, may be difficult to link to affected populations since these effects may not be detectable for years or even generations. By the time an affected population is identified, the chemical may no longer be detectable or even present, making identification of the causitive agent difficult if not nearly impossible. As a result, many retrospective epidemiological studies that associate exposure to an environmental contaminant with human disease have been quite controversial. Such studies emphasize that it is only a matter of time before it becomes clear that we also can be affected by environmental contaminants in much the same way (i.e. endocrine or reproductive problems) as many wildlife species. These two chapters approach ecosystem health in terms of the health of inhabitants of a given ecosystem as do the three previous chapters. In addition, they suggest a broader interpretation, that ecosystem health is related to ecosystem function. In the past many ecosystems were treated as buffers, as if they could absorb anything that was thrown into them, protecting local human populations from their own waste products.

The role of ecosystems as dynamic recycling systems or as buffers is discussed in greater detail in the Chapters 7 and 8 by Matsumura and Cairns respectively. Both of these chapters suggest that ecosystems, as dynamic systems, have some capacity to process, recycle, and break down native or foreign components, and that we are dependent on these processes (e.g. for air quality or water quality) for everyday living. However, there is a limit to what a system can handle. Once we overload the recycling capabilities or buffering capacity we will have altered the ecosystem in such a way that we no longer benefit from the services provided by the 'natural' ecosystems. Neither author attempts to define 'ecosystem health'; in fact Cairns states that we can never really destroy an ecosystem, though we can change it. What might be called a healthy ecosystem is one that carries out the functions we have come to rely upon – forests that assimilate $CO_2$, wetlands that break down and process waste, soil microbes that break down pesticides. How do humans affect these functions? We can alter ecosystems by reducing diversity or by overloading the capacity to recycle, metabolize or renew.

Because of our dependence on ecosystem functions for such basic needs as air quality, the link to human health seems obvious, although it is difficult to 'prove'. As stated by Matsumura 'while the precise statistics and the concrete proofs for the beneficial roles of healthy ecosystems are difficult to find, it is clear that in terms of overall balance, many biological components of ecosystems contribute in reducing the total toxic effects of manmade pollutants'. For example, in industrialized areas where the buffering capacity of many ecosystems has been overwhelmed, there is evidence of increasing human health problems such as increased incidence of lung disease and higher incidence of certain cancers. These two chapters suggest the origin of one definition for ecosystem health: systems that continue to function, change and adapt. The basis for this definition, however, arises from our dependence on such functioning systems for our own health and well-being. It is clear that preserving ecosystem functions and services is beneficial if we want to maintain a certain quality of life. Cairns asserts that we must make fundamental changes in the way we interact with our environment. However, before we truly 'revere and respect life and the environment' we will need to make some very basic changes in our behavior. Changes so basic, according to Cairns, that they will only result from a restructuring of how we think about the environment. The importance of how we think about the environment is discussed in the next set of chapters (Chapters 9–11, by Burger and Gochfeld, Calabrese and Baldwin and Bartell respectively) that deal with sociological and economic values.

As noted by several authors throughout this book, many ecologists frequently ignore the human factor when describing or modeling ecosystems, as though we were not part of earth's ecosystems. However, as discussed in the previous chapters, we can and do affect many ecosystems. Not only have we altered many ecosystems, but we also suffer the consequences of our actions along with what ecologists might consider to be the 'native' inhabitants. If we thought of ourselves as integral parts of an ecosystem, as suggested by Cairns, we might behave in a less destructive manner.

Reducing our impact on the environment most likely will depend on good ecosystem management, which in turn will depend on ecological risk assessment and ecological risk management. Chapters 9–11 provide different views on development of risk assessment techniques.

A risk assessment model already exists for human health. Can this model be adapted or should an analogous model be created for ecological risk assessment? Differences between the two approaches are discussed by Burger and Gochfeld (Chapter 9). According to these authors there are many new problems that must be addressed before a model for ecological risk assessment can be developed. There are some very basic differences.

1. We are no longer dealing with an individual, but with many individuals, many different populations, many different species.
2. The endpoint eludes definition: if human health is a clearly defined endpoint, what is ecosystem health? What do 'we' care about – ecosystem function, stability, adaptability?
3. Most people have fairly similar values when it comes to human health, but whose values matter when determining endpoints for ecosystem assessment?

Burger and Gochfeld propose a combination of risk assessment models that amplify the importance of the target identification phase in ecological risk assessment, a phase that is less critical to human health risk assessment.

Some current ecological risk assessment techniques, for lack of a better approach, use individual indicator species, reducing the complexity of the approach and making the process similar to human risk assessment. The chapter by Calabrese and Baldwin (Chapter 10), however, illustrates a very basic problem with this approach. There can be very great differences in response to toxicants as phylogenetic relatedness is reduced. Although the chapter is written in the context of developing a better uncertainty factor for human risk assessment (since most data are based on toxicity in non-human species, very often rodents), the point is made using a large set of toxicity data in fish. Just between different species of fish, the authors found an approximately 65-fold difference in sensitivity. The greatest difference in classification evaluated was orders-within-class, suggesting that sensitivity differences between lesser related organisms, say vertebrates and invertebrates, are likely to be enormous. The implications are clear; using a single species approach ('protecting' or evaluating a single species, or even a few different species) without regard to the great number of very different species within an ecosystem may result in ineffectual ecosystem risk management.

Rather than completely dismissing the human risk assessment model (based on a 'single [human] species'), Burger and Gochfeld (Chapter 9) offer some additions to the current approach. The authors suggest a revision of the hazard identification phase, which traditionally identifies the substance of concern, the toxicity and the target human population. They assert that when risk assessment techniques are applied to ecosystems the targets and the endpoints require very careful consideration. As discussed above, single targets may be inadequate, and the many biotic components of a system must be considered. The endpoint(s) therefore are much more complex than human health, resulting in the difficulties of making a direct analogy between 'human health' and 'ecosystem health' or even defining 'ecosystem health'.

In the following chapter (Chapter 11), Bartell expands the importance of selecting appropriate endpoints. He suggests that the most basic ecological function is the ability to change or adapt, a concept which can easily be linked to the first chapters in this book, since this most basic function depends on genetic endpoints, DNA and reproduction. Bartell emphasizes, however, that even if we successfully develop techniques for an ecological risk assessment, it will be useless if not incorporated into an ecological risk management framework. Management involves consideration of political and sociological issues. In terms of ecological management, political time scales (such as the lifetime of an elected official or administration) and the political region (local government verses federal) result in 'artificial' temporal scales or boundaries for a given ecosystem. Burger and Gochfeld note that important temporal scales in an ecosystem can range from minutes to decades or centuries, and that spatial scales can be nested (e.g. a pond within a landscape) and can range from small ponds to the global ecosystem. Such concepts do not easily fit into politically driven management schemes.

Finally, Bartell adds an ominous note, asserting that even the best management will be futile if the greatest stress on the environment, human population growth, is not curbed. Population growth and increased consumption of natural resources is a theme that has been indirectly addressed in several earlier chapters, and appears again (as an underlying theme) in the discussion of the economics of ecosystem management by Smith (Chapter 12).

Bartell stresses the importance of environmental management, while Smith and Slovic *et al.* address the difficulties of environmental management (Chapters 12 and 13). Management often comes down to economics, and the assignment of values for ecosystems for use in economic models is a difficult task. Smith (Chapter 12) considers one very basic problem in an economic analysis, how to determine which value system to use when assigning values for ecosystems. Once again this raises several questions – what is ecosystem health, and how important are 'natural' ecosystems, the services provided by an ecosystem, or the diversity of an ecosystem? There is a disparity between economists and ecologists when determining values for ecological resources. Although there have been attempts to adapt either economic models to ecosystems or vice versa, they are based on very different sets of values. While an economist may consider a sustainable economy (at some cost to the environment) as a management goal, an ecologist may consider a sustainable ecosystem (with economic sacrifices) as the goal. There is no solution as to whose values should be applied. Rather than to head in one direction or the other, Smith suggests that a worthwhile goal is to improve information used in policy or management, through improved interdisciplinary research and understanding of the linkages between ecological endpoints and economic commodities. Such information may lead to a better understanding of the possible outcomes of different approaches, which in turn could result in better management decisions.

Slovic *et al.* (Chapter 13) discuss the importance of understanding the 'underlying factors that shape perceptions of environmental risk in various groups' for better risk management. The underlying factors can be influenced by quite disparate backgrounds: e.g. underlying world views, value orientation, prior experiences with nature. Such differences may lead to different perceptions of ecological risk by members of different populations. By identifying what people care about and why, we can begin to understand the relationship between a human population and the environment people live within, which should result in more effective management.

In the end, management decisions that seemed reasonable at one time (such as location of a landfill, or a nuclear power plant or waste water effluent regulations) can result in 'human-made environmental disasters' (e.g. nuclear power plant accidents, landfills that are now superfund sites, contaminated groundwater). Environmental degradation and human health concerns that arise from direct exposure to known levels of radiation or toxic contaminants following a 'human-made' environmental disaster are one issue. Another very important issue is the effect on human health resulting from the *perception* of exposure to environmental contaminants, particularly those that are considered 'human-made' rather than 'natural'. O'Keeffe and Baum (Chapter 14) discuss the psychophysiological consequences of such disasters. Perceptions of individuals exposed to environmental hazards can result in chronic stress. One of the most feared outcomes of exposure (or perceived exposure) is genetic damage (i.e. cancer, or effects on unborn or young children). It is interesting that what may be considered a great threat to humans, altering our DNA and or affecting successful reproduction, parallels what Bartell (Chapter 11) considers to be among the most important ecosystem functions (the ability to adapt and reproduce). Chronic stress (which can be measured in populations up to six years after an accident) in turn may compromise the body's ability to respond to environmental contaminants and can even have lasting effects on the immune system. One component leading to chronic stress according to the authors may be the feeling of a 'loss of control over a system that we created'. The systems created by humans can and have led to environmental contamination and degradation, which in turn have resulted in measurable effects on populations most closely associated with those systems.

Many times in the discussion of these chapters, particularly the chapters addressing risk management, risk perception and economics, the 'human factor' – how we think about the environment – is an underlying issue. It is apparent that if we consider ourselves as integral parts of earth's ecosystems we may care more about our effects on those systems. Many individual populations may think of themselves as separate from 'nature'. Unfortunately, our relationship with a particular ecosystem becomes clear only after the functions or resources we have taken for granted have been affected.

Where did this concept originate? The final chapters of this book, written by Glotfelty and Dixon (Chapters 15 and 16 respectively), literary scholars, attempt to answer this question by reviewing fiction and non-fiction literature of the environment. Glotfelty describes the role of ecocriticism in sorting out the interconnections between the environment and human societies. Two different approaches to literature provide interesting insights into how or why humans interact with their environment. The first approach holds that life reflects literature. This view suggests that the separation of humans from the natural world stems from literature. The most influential example of such literature is the Judeo-Christian bible – in which the earth and its resources were created for humans to use. (If we consider this view, it is not difficult to see why many societies have 'separated' themselves from nature.) The second approach, that literature reflects life, can be used to help understand how people feel about the environment; how individuals within a society view their relationship with the environment. Several examples from this point of view are provided in Dixon's chapter which explores contemporary works of fiction and non-fiction. Ecocriticism and literary analysis, as discussed by Slovic *et al.*, are increasingly important tools for social scientists working to define and understand our perception of the environment.

When we began this project our goal was to open the lines of communication across some very diverse fields of inquiry. We hope this is just a beginning. The topics and content of the chapters were left up to the authors. Some participated in a symposium at the 15th Annual Society for Toxicology and Environmental Chemistry meeting and so had some idea of what other contributors were thinking about, but many of the authors were not present. Regardless, there were many similarities in the underlying themes of the chapters as discussed above. The basic theme to any chapter was that there are links between degradation of the environment and human health. The links can be direct, e.g. exposure to environmental contaminants such as endocrine disrupters, or indirect, e.g. reduced water quality as a result of wetlands destruction.

The questions raised by the title of this book, and by many of the authors in it are: can contaminated ecosystems, or ecosystems that are destroyed as a result of human activity, be referred to in terms of ecosystem health? What is ecosystem health? After struggling with definitions for 'ecosystem health', it has become clear through many of these chapters that there are some very good reasons why it is a difficult term to use. When we talk about human health, there are often boundaries, individuals; when we talk about ecosystems the boundaries are arbitrary. We have a pretty good idea of the range of 'proper' functioning for a human, but this is not so clear for an ecosystem. The values for determining human health depend on individual values, and most people are in agreement about what constitutes a healthy body and mind. Whose values should be applied when evaluating an ecosystem?

In concurrence with statements by many contributors to this book, we hope that this book serves to open lines of communication across diverse disciplines and perspectives, including natural scientists, social scientists, writers and critics, and the 'lay' public concerned with environmental management. Given the degrees of specialization that have evolved within fields of inquiry, we recognize that this is a daunting task. Cairns (Chapter 8) notes that: 'Too much emphasis on specialization decreases interactions with other disciplines, which in turn diminishes any effectiveness in coping with problems that transcend the capabilities of a single discipline (i.e. all of human society's major problems).' Clearly, the subject of connectedness between the 'healthful' structure and function of humans and ecosystems falls into this domain.

# Index

GPSR Compliance

*The European Union's (EU) General Product Safety Regulation (GPSR) is a set of rules that requires consumer products to be safe and our obligations to ensure this.*

*If you have any concerns about our products, you can contact us on ProductSafety@springernature.com*

In case Publisher is established outside the EU, the EU authorized representative is:

Springer Nature Customer Service Center GmbH
Europaplatz 3
69115 Heidelberg, Germany

**Batch number: 10761599**

Printed by Printforce, the Netherlands